RAND

# *Calculating the Utility of Attacks Against Ballistic Missile Transporter-Erector-Launchers*

*Russell D. Shaver, Richard F. Mesic*

*Prepared for the United States Air Force*

**Project AIR FORCE**

Approved for public release; distribution unlimited

# PREFACE

The motivation of this analysis, performed in support of the Air Force's efforts in theater missile defense (TMD), was to bring increased attention to the development and fielding of capabilities to hold at risk enemy theater ballistic missiles (TBMs) and their transporter-erector-launchers (TELs) through attacks on the TELs either before or after TBM launch.

The report should be of interest to analysts in the TMD community and to others interested generally in the application of probabilistic and Monte Carlo methods to military problems.

The study was sponsored by the Air Combat Command Armament Requirements Division (ACC/DRA) and took place within the Force Modernization and Employment Program of Project AIR FORCE, the Air Force's federally funded research and development center at RAND.

## PROJECT AIR FORCE

Project AIR FORCE, a division of RAND, is the Air Force federally funded research and development center (FFRDC) for studies and analyses. It provides the Air Force with independent analyses of policy alternatives affecting the development, employment, combat readiness, and support of current and future aerospace forces. Research is being performed in three programs: Strategy, Doctrine, and Force Structure; Force Modernization and Employment; and Resource Management and System Acquisition.

Project AIR FORCE is operated under Contract F49620-91-C-0003 between the Air Force and RAND.

Brent Bradley is Vice President and Director of Project AIR FORCE. Those interested in further information concerning Project AIR FORCE should contact his office directly:

Brent Bradley
RAND
1700 Main Street
P.O. Box 2138
Santa Monica, CA 90407-2138

## CONTENTS

Appendix

## FIGURES

# TABLES

## SUMMARY

Under the assumption that future opponents will choose to acquire inventories of theater ballistic missiles (TBMs) substantially larger than the number of their transporter-erector-launchers (TELs),[1] this report supports the view that counterforce operations against the TELs can play a role in sharply reducing the overall size of a prospective threat. For initial inventory ratios of 10 TBMs per TEL (a ratio not inconsistent with past practice by many countries), reductions of approximately 80 percent in missiles launched are possible with probabilities of successful postlaunch TEL kill of about 0.5. Even for probabilities of TEL kill of only 0.2, reductions of 50 percent are possible. Combined prelaunch and postlaunch counterforce attacks act synergistically, enhancing the overall effectiveness. The report does not discuss the circumstances in which these attacks can achieve any specific degree of effectiveness. History suggests that claims for significant counterforce capabilities should be viewed with skepticism. While we concur with this observation, there may exist a significant motivation for striving to make this capability at least modestly effective. We believe that this is clearly possible for postlaunch counterbattery operations.

It is useful to note that postlaunch counterforce attacks (often called counterbattery attacks) do little initially to affect TBM launches. As

[1]This assumption is reasonable for current threats because the cost of the TELs is significantly greater than the cost of the TBMs. In the future, as missiles get smarter and TELs therefore can become less complex, the TELs may become single-shot "throw-aways." For the foreseeable future, however, we would expect to see many reloads per TEL, although not necessarily many tons, as was the case in Iraq.

such, counterbattery capabilities are properly considered to be adjuncts to various active defense measures for defeating TBM threats. They aid these active defensive measures mainly by lessening the magnitude of the attack (both in terms of the total number of TBMs that can be launched and their launch rate), thus potentially lowering the total investment in active defenses.

The report also provides the reader with a set of equations (and their derivations) to calculate additional outcomes. For those interested in such calculations, an appendix lists computer programs suitable for this purpose.

# ACKNOWLEDGMENTS

The authors benefited greatly from discussions with other RAND staff in developing these ideas. In particular, we wish to thank David Vaughan for early contributions that helped motivate this work and for his careful review of an early draft. Additional thanks go to Keith Henry for his thoughtful review and constructive comments.

Chapter One

# INTRODUCTION

The purpose of this report is to bring increased attention to the development and fielding of capabilities to hold at risk enemy theater ballistic missiles (TBMs) and their transporter-erector-launchers (TELs) through attacks on the TELs either before or after TBM launch. The "conventional wisdom" seems to be that postlaunch (or "counterbattery") attacks on the TELs are feasible (because of the exploitable missile launch signature), but not very useful. After all, the process starts with the *successful* launch of the threat missile, and empty TELs have little intrinsic value in themselves. On the other hand, successful prelaunch attacks against the TBMs/TELs would be highly useful (killing the TBMs before they were launched, thereby avoiding a host of end-game concerns), but are regarded as technically and operationally far more challenging and thus have an extremely low probability of success. Thus, conventional wisdom further suggests that efforts in both areas are likely to have low payoff. This study challenges that assessment. We do not dispute that these concepts are challenging, although we do agree that counterbattery attacks are inherently easier to achieve than prelaunch kills. But we do believe that even modest capabilities could be very useful. If those capabilities could be fielded with relatively little expense (as we believe is the case by adapting existing systems such as the F-15E Strike Eagle), then our analysis supports the view that they could be "best buys."[1]

[1]This statement assumes that the United States has control of the airspace over potential TBM operating areas. Assuming this to be the case, the additional option of boost-phase kill also exists. The synergy between prelaunch, boost-phase, and

In recent work by David Vaughan and Richard Mesic, the value of a counterbattery attack was represented by the number of residual TBMs that could no longer be launched if TELs were located and destroyed immediately after their associated TBMs were launched.[2] Figure 1 shows the results of a calculation that was used to describe this result. In this simplistic calculation it was assumed that there were an infinite number of potential reloads, so that the average number of successful launches could be derived from the geometric probability distribution (i.e., the expected number is 1/postlaunch TEL kill probability). In Figure 1 we simply truncated these curves at

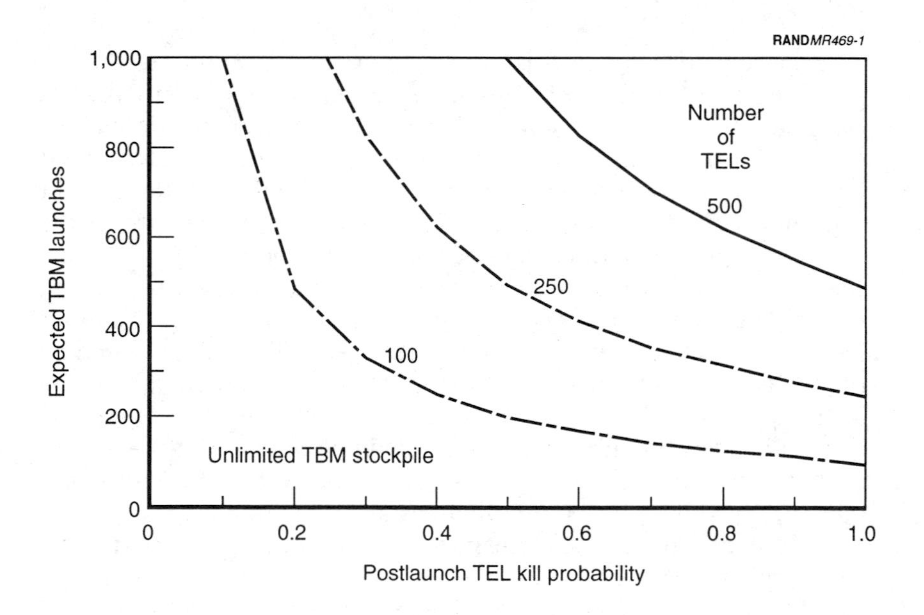


**Figure 1—Counterbattery Attacks Can Ground Reload Missiles**

---

postlaunch attacks against the missile and the TEL helps make all three more cost-effective.

[2]Unpublished work by David Vaughan and Richard Mesic on investment strategy for theater missile defense; unpublished briefing by Tim Naff (U.S. Army Space and Strategic Defense Command [SSDC]), "TMD Attack Operations and Active Defense Functional/Quantitative Relationships."

the assumed total TBM inventory of 1000. The Naff briefing made the point that even relatively small single-engagement kill probabilities could substantially reduce the total size of the attack by killing all the TELs before all the TBMs could be launched. While a full complement of active terminal-area defenses would still be required to handle the TBMs that were launched, the total attack against those defenses would be reduced.[3]

The Naff briefing supported the feasibility of obtaining a competent counterbattery capability. Not only would the exhaust plume of the in-flight TBM provide an excellent cue for alerting the searching sensors to the presence of a TEL, but backtracking the missile's trajectory to locate a launch point on the ground would provide a relatively small footprint area within which the TEL could be located. Various practical operational concepts exist for prosecuting the cue and the backtrack into a successful attack against the empty TEL, including the employment of F-15Es with suitable software modifications to the APG-70 radar.[4]

Thus, Figure 1 supports arguments in favor of obtaining a robust counterbattery capability. However, it begs some additional questions:

- Figure 1 is based on an approximation (i.e., it assumes an infinite supply of TBMs). Would more exact (hence, more complex) analyses such as we develop here for more realistic inventory sizes and operational constraints (e.g., TEL and TBM clustering and finite numbers of TBMs) show a reduced utility for counterbattery attacks?
- How would the results change if prelaunch counterforce attacks were included? Although the actual capability of prelaunch at-

---

[3]The reduction in threat size is sensitive to the ratio of the attacker's TBMs to TELs. The greater the ratio, the higher is the utility of postlaunch counterbattery attacks. While the ratio of TBMs to TELs is obviously within the control of the opponent, historically countries have chosen many more TBMs than TELs. Ratios of 10 or larger are the rule.

[4]See, for example, Mesic and Vaughan, "A Concept for Near-Term Action on TMD Counterforce Capabilities Development and Demonstration," May 1993, and Mesic, Vaughan, and Shaver, "Theater Missile Defense (TMD) Pillar Balancing: A Strawman Approach and Initial Observations," April 1993 (project memoranda to the Air Force).

tacks might be significantly less than postlaunch, TELs would be killed in both.

- Given the random nature of the outcomes, with what confidence could we assert a particular outcome, given that we knew (or were willing to guess at) the likelihood of successful location, identification, and kill? If counterbattery attacks rest on killing all the TELs, the tails in the distribution of outcomes could be extremely important.

We address these questions in an analytic context. The following chapters will first cover counterbattery attacks when only a single TEL is involved, then extend the analysis to cover multiple TELs deployed as part of a cluster of TBMs and TELs, and then further extend the analysis to include prelaunch counterforce attacks. A few observations follow. Appendix A contains supporting mathematics that should help anybody wishing to pursue these subjects. Appendix B lists various computer programs (written in QuickBASIC) that were developed to support the figures shown in this report.

# Chapter Two

# THE SIMPLE MATHEMATICS FOR COUNTERBATTERY ATTACK OUTCOMES ASSUMING A SINGLE TEL

Assume that a single TEL has access to N TBMs—one on the TEL itself and an additional N – 1 back at the resupply site. Further assume that if the TEL is killed, whatever TBMs remain at the resupply site can no longer be launched (later we will assume that other TELs can access these TBMs, an assumption that will affect our outcomes). Then the probability that a TEL successfully launches exactly K TBMs is simply the probability that it survives the counterbattery attack after the first K – 1 launches but is killed after the Kth launch, that is,

$$P(K) = \left(1 - P_k\right)^{K-1} P_k \qquad 1 \le K < N \qquad (1)$$

and

$$P(N) = \left(1 - P_k\right)^{N-1} \qquad K = N \qquad (2)$$

where $P_k$ is the probability that the TEL is killed after any particular launch.

Using Eqs. (1) and (2), we can derive the following formula for the expected number of TBMs launched, E(K):

$$E_1(K) = \frac{1 - \left(1 - P_k\right)^N}{P_k} \qquad (3)$$

where the numerical subscript after E identifies the size of the TEL inventory (in this case, one). The derivation of Eq. (3) can be found in Appendix A. For very large N or very small TEL, survival probability (for small values of $(1 - P_k)^N$), $E_1(K)$ becomes simply

$$\frac{1}{P_k},$$

the approximated formula used in the earlier Vaughan-Mesic work. In addition, if we let $P_k$ approach zero, then $E_1(K)$ approaches N; that is, if the TELs are never killed, they launch the full inventory of TBMs. Figure 2 plots the expected size of the TBM inventory launched, $E_1(K)$, as a function of $P_k$ for various TBM inventory sizes. Shown for comparison is the expected number of TBMs launched, assuming an infinite supply of TBMs, so that the nature of the errors in that simple approximation can be seen. As the counterbattery kill potential falls, the divergence between the finite and infinite TBM

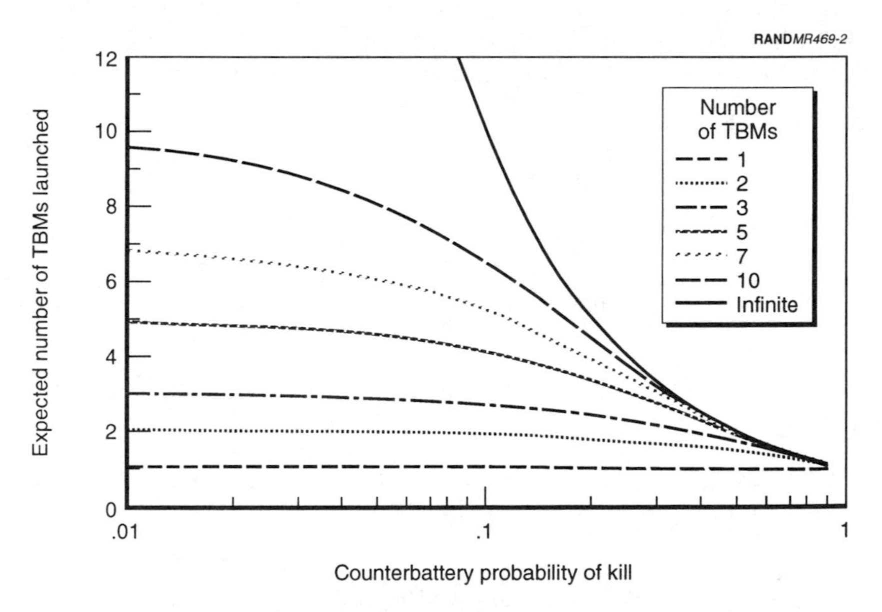


**Figure 2—Ability of Counterbattery Attacks to Limit the Expected Number of TBMs Launched, Assuming a Single TEL**

stockpiles grows, with the finite stockpile outcome asymptoting to its inventory size as $P_k$ approaches zero.

Two points are worth noting:

- For TBM inventories per TEL greater than about 5, counterbattery $P_k$s as modest as 0.2 can dramatically reduce the total fraction of the TBM inventory that can be launched. Other work suggests that $P_k$s of 0.2 or higher (perhaps much higher) may be achievable at only modest cost and risk.[1]
- High counterbattery kill potential can limit the number of TBMs launched even if the inventory per TEL is relatively small. Of course, if there is only one TBM per TEL, then counterbattery attacks have no analytic effect (although the effect on the morale and motivation of the doomed launch crews might be significant).

Because of the random nature of these outcomes, it is also of interest to look at the variance of these outcomes. Figure 3 plots the statistical variance (the square of the standard deviation) of these outcomes as a function of $P_k$ for several values of N. Note that the variance can be relatively large at intermediate values for $P_k$ (peaking where $P_k$ = 1/the number of TELs), leading to concerns that simply using expected values for counterbattery attack outcomes in multilayer attack calculations may not properly capture the consequences of unlucky occurrences. For example, for $P_k$ = 0.1 and 10 TBMs per TEL, the expected number of launches (from Figure 2) is about 6.5, but the standard deviation (the square root of the curves in Figure 3) is about 4—a significant number compared with the expected value.

---

[1]See, for example, Richard Mesic, "Extended Counterforce Options for Coping with Tactical Ballistic Missiles," in Paul Davis (ed.), *New Challenges for Defense Planning*, RAND, Santa Monica, CA, 1994.

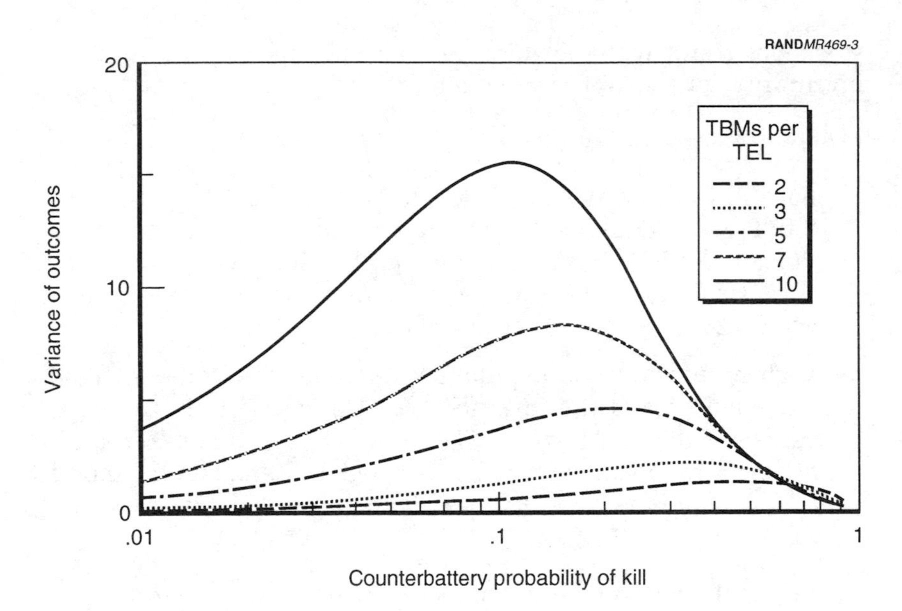


**Figure 3—Variance of Counterbattery Attack Outcomes for 2–10 TBMs and a Single TEL**

Chapter Three

# EXTENDING THE OUTCOMES TO CASES WHERE MULTIPLE TELs HAVE ACCESS TO A COMMON CACHE OF TBMs

The simple model of a single TEL with TBM reloads that it alone can access and launch is probably not a realistic portrayal of how most enemies would operate. More likely is a model of a cache of TBMs and a number of TELs that have access to that cache. Unless all TELs with access to that cache are killed, the remaining TBMs in the cache can still be launched.

The mathematics associated with calculating expected outcomes in cases where TBMs and TELs are clustered into caches of multiple TELs (T) and TBMs (N) is provided in Appendix A. No simple closed-form solutions equivalent to Eq. (3) exist, and solving the equations must be done numerically on computers. Three approaches were used. The first approach yields a set of algebraic equations that, while not reducible to a simple closed form, can be easily calculated by numerical means. This approach provides exact answers for both the expected values of the outcomes and the associated distribution functions for the number of TBMs actually launched for any specific set of inputs. It also provides the reader with an understanding of how the primary input variables affect the outcome without performing extensive calculations. However, the numerical calculations can be time-consuming if the number of TELs in a cluster is large (greater than 5), and thus is only useful for small T.

A second approach is to computationally solve a matrix calculation patterned after Markov. This approach also produces exact answers for both the expected values of the outcomes and the associated distribution functions for the number of TBMs actually launched for any specific set of inputs. It is not computationally limited to small numbers of TELs; it can rapidly calculate outcomes for essentially any threat size.

The third and last approach uses a Monte Carlo computer simulation to yield estimates (rather than exact calculations) of expected outcomes and data on outcome distributions; it also works for large threat sizes. If the number of trials is sufficiently large, the uncertainties in outcome estimates resulting from the stochastic character of the approach can be minimized.

Table 1 summarizes these characteristics. Appendix B contains the computer codes developed for each of these approaches.

The next several figures address the question of the importance of assuming that TELs and TBMs are clustered into sets wherein every TEL has access to and can launch every TBM. It is possible for there to be multiple clusters, in which case we assume that a TEL from one cluster cannot launch TBMs that belong to another (if they could, then we would simply increase the size of the cluster). An extreme case is if all TELs have access to all TBMs. For practical reasons, this seems unlikely. However, the other extreme—each TEL can access only its specifically assigned reload TBMs (i.e., all clusters consist of single TELs, as mathematically portrayed in Chapter Two)—seems equally unlikely.

Figure 4 starts with the assumption that the total number of TBMs is ten times the total number of TELs. It shows how clustering TELs diminishes the effectiveness of postlaunch counterbattery attacks, normalizing the outcome to the TBM/TEL ratio.

**Table 1**

**Mathematical Approaches to Calculating Counterbattery Attack Outcomes**

| Calculation | Calculated Outcomes | | Computer |
|---|---|---|---|
| Method | Expected Values | Distributions | Limitations |
| Algebraic | Exact | Exact | Requires small number (T) of TELs for cluster |
| Markov matrix | Exact | Exact | None |
| Monte Carlo | Approximate | Approximate | None |

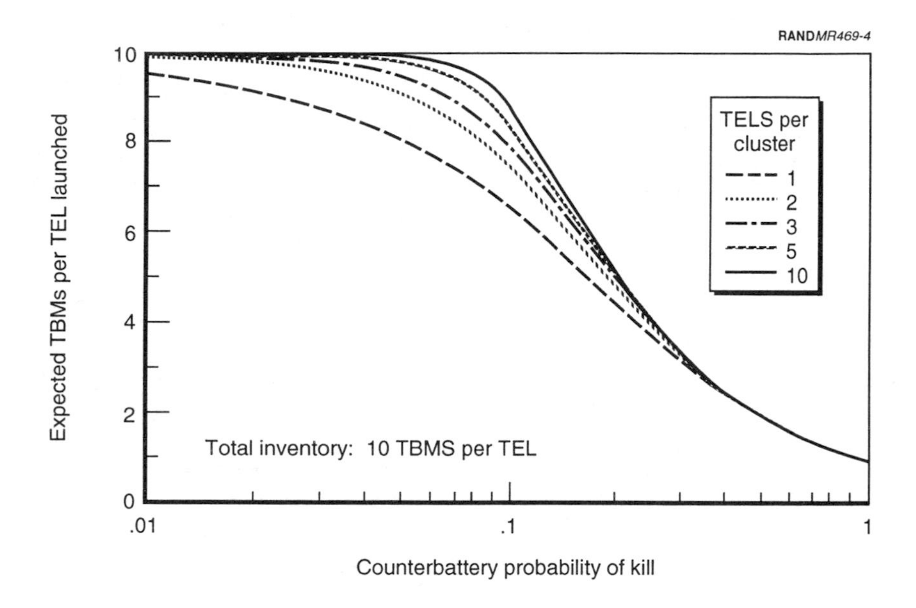


**Figure 4—Expected Number of TBMs Launched (Normalized to Total Number of TELs) as a Function of the Number of TELs in a Cluster**

Several observations can be drawn from Figure 4:

- If the counterbattery kill probability is above about 0.2, then the clustering of TELs does not significantly affect the outcome.
- Clustering ratios above about 5 have diminishing returns.
- Regardless of clustering, counterbattery kill probabilities above about 0.2 can still sharply diminish the total number of TBMs that can be launched.

Another important variable is the ratio of total TBMs to TELs. Assuming a cluster of 100 TBMs, Figure 5 shows the expected number of these TBMs that could be launched as a function of the number of TELs added to the cluster. We see the following:

- An obvious and effective counter to counterbattery attacks is to buy additional TELs. The sensitivity is substantial, but even

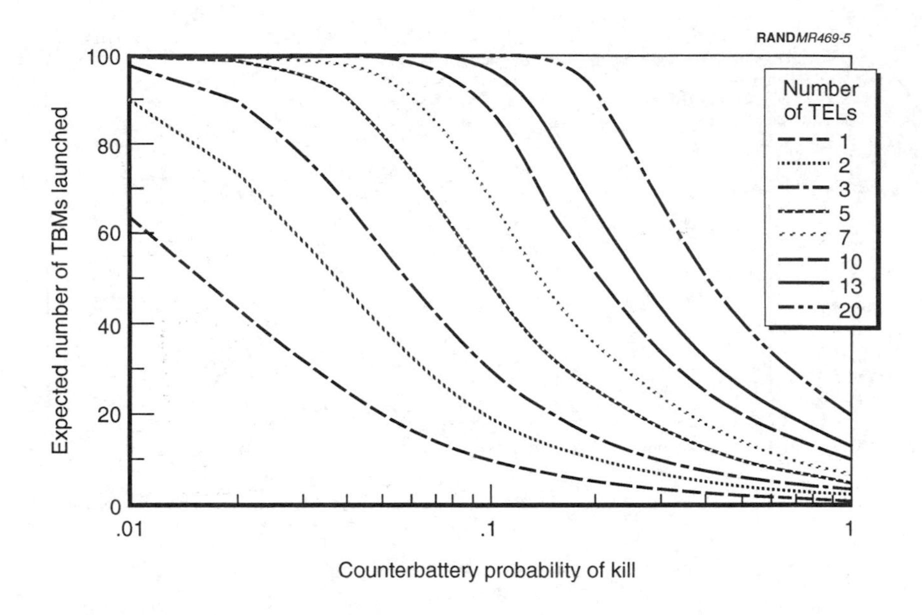


**Figure 5—Sensitivity of Counterbattery Attack Effectiveness to TEL Proliferation**

buying a fairly large number of TELs does not totally obviate counterbattery effectiveness.

- If $P_k$'s of 0.5 or higher are achievable, even a large number of TELs cannot guarantee that the majority of TBMs in the inventory can be successfully launched.

It is impossible, of course, to judge from these calculations whether TEL proliferation constitutes a cost-effective response to a counterbattery threat. Past experience with regard to TBM deployments suggests that countries have decided the issue by buying a large number of TBMs per TEL. No doubt this decision was influenced in part by the perception that future opponents could not mount successful counterbattery attacks. But it almost certainly was also affected by peacetime operating costs and other constraints.

Chapter Four

# AN EXAMPLE ANALYSIS, ASSUMING 5 TELs AND 50 TBMs IN A CLUSTER

The following example demonstrates how an analysis of counterbattery operations might be carried out.

Assume that the threat consists of an enemy who has 500 TBMs and 50 TELs, divided into 10 clusters. Each cluster is assumed to operate autonomously. The questions we address are (1) on the average, how much would counterbattery attacks reduce the overall size of the threat? (2) with what confidence can we state these outcomes? and (3) how sensitive are the answers to assumptions about how the enemy operates and the probability that counterbattery attacks will succeed?

Figure 6 shows the expected number of TBMs that would be launched from a specific cluster as a function of counterbattery effectiveness; it simply repeats one of the curves in both Figures 2 and 4. It could have been calculated by any of the three approaches, although the first two give an exact answer. For comparison, we have included the infinite TBM case. As noted, counterbattery kill probabilities above about 0.2 result in a strong likelihood that the TELs would be killed prior to TBM exhaustion, thus sharply curtailing the total number of TBMs that would be launched.

Also shown in Figure 6 are the 10 percent and 90 percent outcomes. The lower boundary is defined by specifying that 90 percent of the outcomes exceed it, whereas only 10 percent exceed the upper boundary. The curves show that outcomes can vary widely.

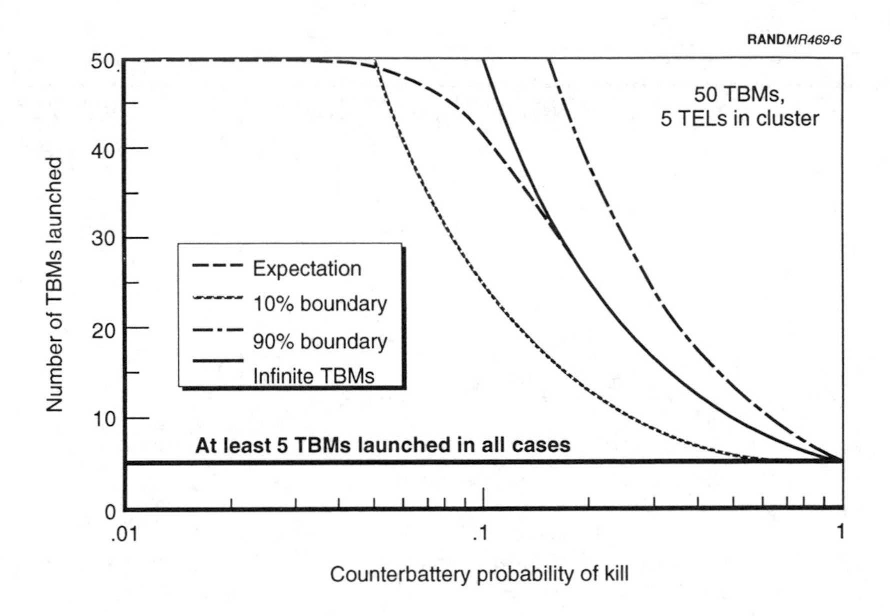


**Figure 6—Expected Values Versus Upper and Lower Boundaries, 50 TBMs and 5 TELs**

If we had constructed similar curves for the full inventory of 500 TBMs, the boundaries would have become narrower. We have not calculated the results, but they can be approximated by adjusting the differences by

$$\sqrt{10},$$

where 10 comes from the ratio of the total inventory to the number of TBMs in the cluster.[1]

Figure 7 is shows the distribution of outcomes for this case another way. This curve has been calculated for $P_k = 0.2$. Note that the

[1]Assuming ample TELs, the variance of the results about the mean, normalized by the mean, go as $1/\sqrt{n}$, where n is the total number of TBMs. As TEL survival comes into play, the variance will change. However, assuming a factor of 10 in increased numbers of both TBMs and TELs, the normalized variance of the larger inventory should be about $1/\sqrt{10}$.

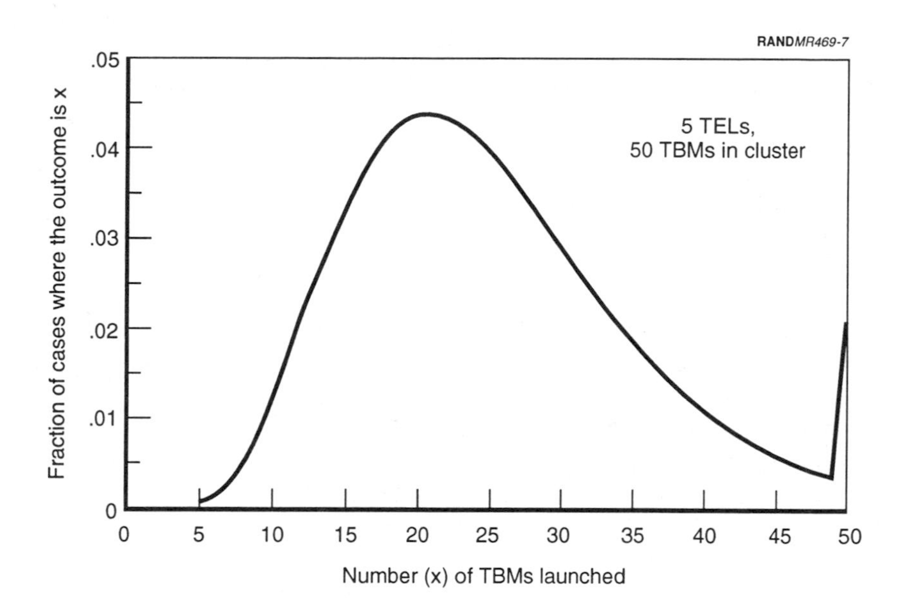


**Figure 7—Fraction of Cases Where Specific Number of TBMs Is Launched, $P_k = 0.2$**

maximum likelihood outcome is about 20 TBMs, although the mean is just under 25. The "jump" or spike at the end corresponds to the fraction of trials where all of the TBMs are launched. This spike is large whenever the counterbattery $P_k$ is sufficiently small that multiple TELs survive to TBM exhaustion. Note that the spread in the outcome is substantial, as we could have anticipated from Figure 6.

For comparison, we also ran the Monte Carlo model for this set of inputs. Table 2 shows the outcomes of the run. The numbers in the histograms are the occurrences where the given number of TBM launches or surviving TELs was the result (the number of TBMs launched can run from a minimum of 5 to a maximum of 50; the surviving TELs from 0 to 5). For example, out of the 500 trials run for this case, in 482 all the TELs were killed, and in 20 trials exactly 26 TBMs were launched (read '6' across the top and '20' down the side to find the number of cases where 26 TBMs were launched). For this specific run, the calculated average number of TBMs fired is slightly

## Table 2

## Output from Monte Carlo Model (Counterbattery Attacks) I

### Input Data

Total TELs in cluster = **5**
Total TBMs in cluster = **50**
$P_k1$ = **0**
$P_k2$ = **0.2**
Number of trials = **500**

### Output Data

Average number of TBMs fired = **26.136**
Standard deviation = **10.109**
Standard deviation of mean estimate = **0.452**

### Histogram of Number of TBMs Fired

| Tens | Ones Count | | | | | | | | | |
|---|---|---|---|---|---|---|---|---|---|---|
| Count | 0 | 1 | 2 | 3 | 4 | 5 | 6 | 7 | 8 | 9 |
| **0** | | | | | | 0 | 0 | 3 | 2 | 0 |
| **10** | 6 | 8 | 8 | 9 | 17 | 16 | 16 | 20 | 23 | 20 |
| **20** | 22 | 20 | 21 | 16 | 17 | 28 | 20 | 12 | 20 | 12 |
| **30** | 8 | 16 | 14 | 10 | 9 | 17 | 14 | 5 | 8 | 4 |
| **40** | 8 | 1 | 9 | 6 | 2 | 4 | 2 | 5 | 2 | 0 |
| **50** | 20 | | | | | | | | | |

### Histogram of Number of TELs Surviving

| Tens | Ones Count | | | | | |
|---|---|---|---|---|---|---|
| Count | 0 | 1 | 2 | 3 | 4 | 5 |
| 0 | 482 | 15 | 2 | 0 | 1 | 0 |

higher than the theoretical expected outcome (in fact, given the standard deviation of the mean of the estimate, this case is a substantial outlier). Figure 8 plots the TBM data in Table 2 on top of the data in Figure 7.

What can we derive from this? In some plausible scenarios, a reduction in both the total size of the threat and in the salvo sizes possible at any particular time could greatly enhance TMD effectiveness; in other scenarios the effects might be small. Certainly, counterbattery attacks of this character do little to deny a threat based on weapons of mass destruction, where denial of *any* successful TBM penetrations is the dominant criterion. To deal with such threats, we need to consider both postlaunch active defense options (e.g., Theater High-Altitude Area Defense [THAAD] or boost-phase defenses) and prelaunch counterforce options.

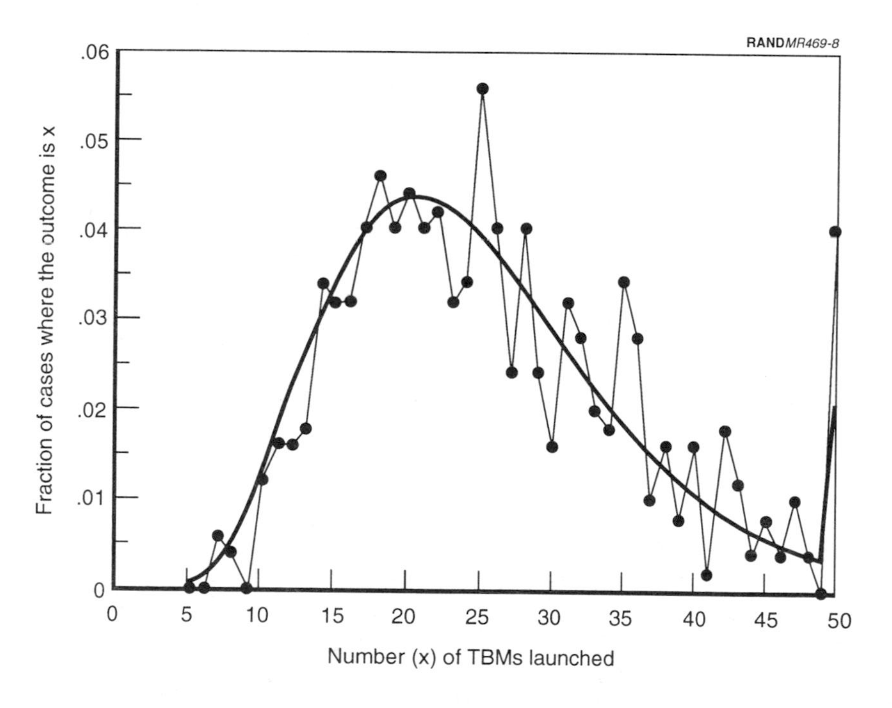


**Figure 8—Monte Carlo Outcomes Superimposed on Figure 7, $P_k = 0.2$**

Chapter Five

# EXTENDING THE CALCULATIONS TO INCLUDE PRELAUNCH COUNTERFORCE ATTACKS

The above calculations presumed that the only threat to the TELs and TBMs came from attacks against empty TELs after the TBM had been launched. It is obvious that the same aircraft that might execute the postlaunch counterbattery attack will be in a position to carry out prelaunch counterforce attacks if the TBM/TEL can be found prior to launch. The mathematics for prelaunch counterforce attacks can be found in Appendix A. We will not repeat the math here. However, for comparison with Eq. (3), we offer the general formula for the expected number of TBMs launched assuming a single TEL.

$$E_1(K) = \frac{\left(1-{}^1P_k\right)}{{}^1P_k} \cdot \left(1 - \left(1-{}^1P_k\right)^N\right) \qquad (5)$$

where we have designated the prelaunch single-engagement kill probability as ${}^1P_k$ to differentiate it from the engagement kill probability associated with postlaunch attacks (we will designate postlaunch single-engagement kill probabilities by ${}^2P_k$). A comparison with Eq. (3) shows that the two formulas are identical with the exception of the factor

$$\left(1-{}^1P_k\right)$$

in the numerator. Note that when the probability of TEL survival over N launches is very small (i.e., when

$$\left(1-{}^{1}P_k\right)^N$$

is small), Eq. (5) shows that, as is intuitively obvious, one less launch would be expected than in the postlaunch case in Eq. (3).

Figure 9 is equivalent to Figure 2, except it assumes that there are only prelaunch counterforce attacks against a single TEL. Note that these curves are slightly more favorable to the attacker, because a kill results in denying a TBM launch as well as killing the TEL.

Figure 9 depicts the no TEL clustering case. Figure 10, the equivalent of Figure 4, shows the impact of TBM/TEL clustering on the prelaunch counterforce outcome. Again, the similarity to Figure 4 is not surprising, with counterforce attacks of equal kill potential yielding slightly more favorable results for the attacker. However, we note that prelaunch counterforce attacks are not blessed with the same certainty of a cueing signal as are the postlaunch counter-

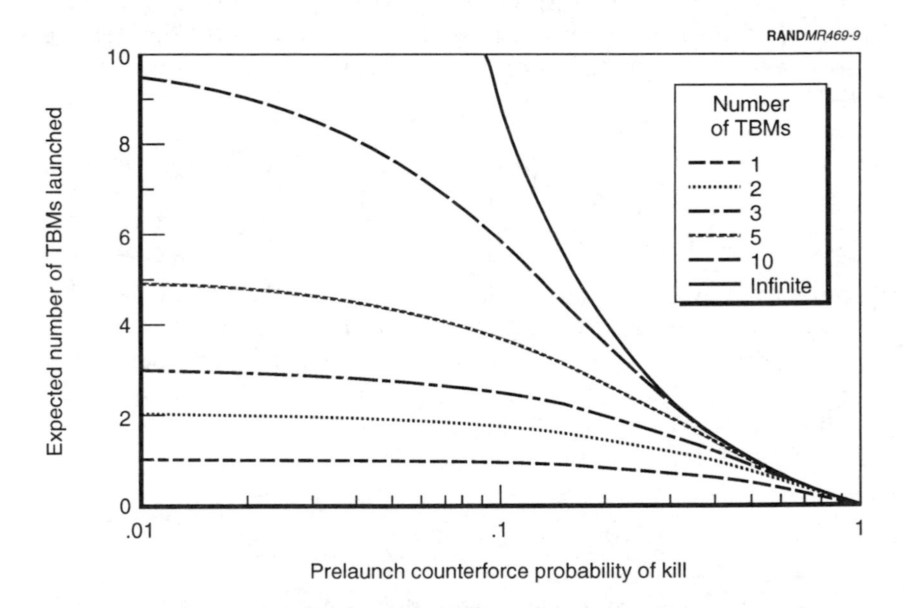


**Figure 9—Ability of Prelaunch Counterforce Attacks to Limit the Expected Number of TBMs Launched, Assuming a Single TEL**

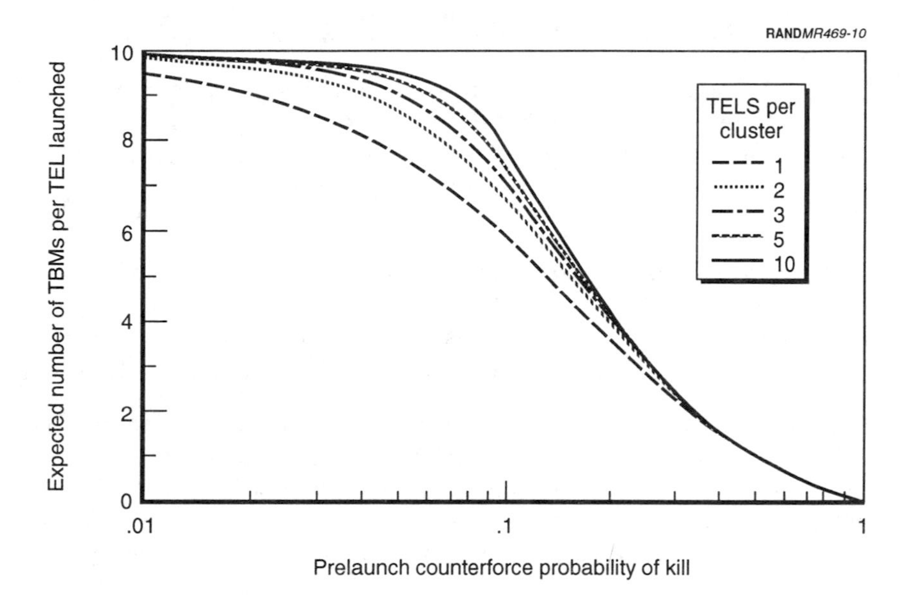


**Figure 10—Effect of Clustering on Prelaunch Counterforce Attack Outcomes**

battery attacks. In the latter case, the bright infrared (IR) signal associated with TBM launch is virtually impossible to mask or otherwise conceal. Finding TELs without such a cue proved to be almost impossible during the Gulf War, and it is not certain that significant progress has been made since then to reverse this unfavorable outcome.

*Combined* prelaunch and postlaunch counterforce attacks would obviously improve the outcomes. The attacker would get two opportunities to kill the TEL during a launch sequence, effectively increasing the kill probability as well as killing TBMs if the prelaunch attack succeeds. The expected number of TBMs launched for a single TEL—the equivalent equation to Eqs. (3) and (5)—is simply

$$E_1(N) = \frac{\left(1-{}^1P_k\right)\left(1-\left[\left(1-{}^1P_k\right)\left(1-{}^2P_k\right)\right]^N\right)}{1-\left(1-{}^1P_k\right)\left(1-{}^2P_k\right)} \quad (6)$$

where the prescripts 1 and 2 denote prelaunch and postlaunch.

Figure 11 displays outcomes from combined pre- and postlaunch attacks, where it is assumed that the cluster consists of 100 TBMs and 10 TELs. We have parameterized the postlaunch attack probability, plotting curves for different prelaunch capabilities.

The results in Figure 11 show a striking synergy between prelaunch and postlaunch attack effectiveness. Even modest prelaunch capabilities (e.g., $P_k$s in the neighborhood of 0.2) can significantly augment the effectiveness of postlaunch counterbattery attacks, reducing the number of TBMs that can be launched by 50 percent or more.

Clearly, the countermeasure of adding more TELs would lessen the effectiveness of the combined attacks. In general, however, combined attacks can sharply curtail the total launch potential of a TBM-armed enemy. In addition, counterforce attacks, particularly prelaunch attacks, will aid theater missile defenses by lessening the number of TBMs launched during any launch period as well as overall. Counterforce attacks in combination with boost-phase

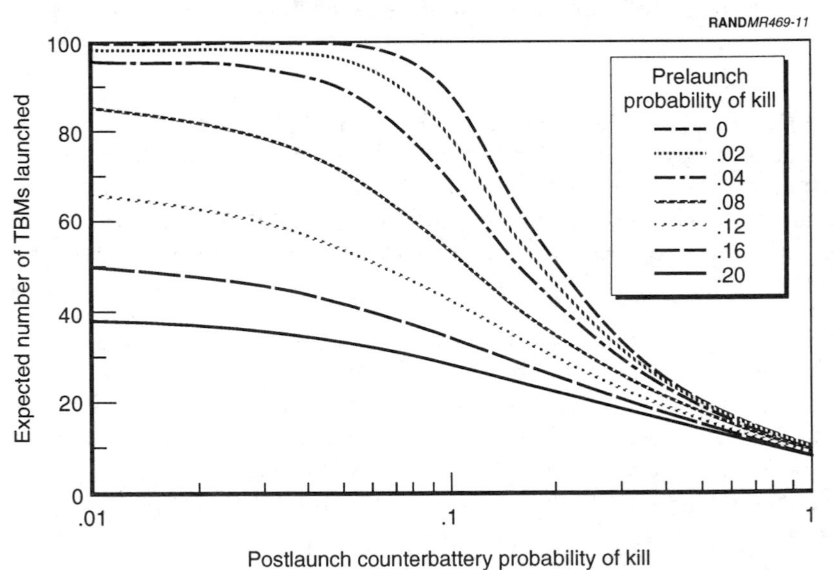


**Figure 11—Combined Effectiveness of Prelaunch and Postlaunch Counterforce Attacks: 100 TBMs and 10 TELs in Cluster**

attacks, attacks that could be carried out by the same aircraft, would make any conceivable enemy countermeasures to most theater missile defense architectures much less effective. Thus, counterforce attacks can directly (as well as indirectly) support overall theater missile defense architectures, improving the effectiveness of other components of the defense.

We return to the earlier example of 50 TBMs and 5 TELs. Table 3 repeats Table 2, with an overlay of a prelaunch counterforce attack effectiveness of 0.1.

**Table 3**

**Output from Monte Carlo Model (Counterbattery Attacks) II**

## Input Data

Total TELs in cluster = **5**
Total TBMs in cluster = **50**
$P_k1$ = **0.10**
$P_k2$ = **0.20**
Number of trials = **500**

## Output Data

Average number of TBMs fired = **16.378**
Standard deviation = **7.185**
Standard deviation of mean estimate = **0.321**

### Histogram of Number of TBMs Fired

| Tens Count | Ones Count | | | | | | | | | |
|---|---|---|---|---|---|---|---|---|---|---|
| | 0 | 1 | 2 | 3 | 4 | 5 | 6 | 7 | 8 | 9 |
| 0 | 0 | 0 | 2 | 1 | 3 | 4 | 7 | 14 | 17 | 18 |
| 10 | 35 | 40 | 25 | 35 | 39 | 29 | 23 | 22 | 20 | 25 |
| 20 | 16 | 23 | 16 | 14 | 10 | 8 | 8 | 4 | 4 | 7 |
| 30 | 6 | 2 | 4 | 3 | 1 | 3 | 3 | 3 | 4 | 1 |
| 40 | 0 | 0 | 0 | 1 | 0 | 0 | 0 | 0 | 0 | 0 |
| 50 | 0 | | | | | | | | | |

### Histogram of Number of TELs Surviving

| Tens Count | Ones Count | | | | | |
|---|---|---|---|---|---|---|
| | 0 | 1 | 2 | 3 | 4 | 5 |
| 0 | 500 | 0 | 0 | 0 | 0 | 0 |

Chapter Six

# OBSERVATIONS

Representative calculations of the potential implications of postlaunch and prelaunch counterforce attack operations against TELs and their associated TBMs indicate that counterforce operations can substantially reduce the overall threat of TBM attacks. It can achieve this effect even if the probabilities of success for individual attacks are modest. For example, with initial inventory ratios of 10 TBMs per TEL (a ratio not inconsistent with past practice by many countries), reductions of approximately 80 percent in missiles launched are possible with probabilities of successful postlaunch TEL kill of about 0.5. Even for probabilities of TEL kill of only 0.2, reductions of 50 percent are possible. Combined prelaunch and postlaunch counterforce attacks act synergistically, enhancing the overall effectiveness. We have not, however, discussed the circumstances in which these attacks can achieve any degree of effectiveness. History suggests that claims for significant counterforce capabilities should be viewed with skepticism. While we concur with this observation, this discussion suggests that there exists a significant motivation for striving to make this capability at least modestly effective. We believe that this is possible for postlaunch counterbattery operations.

Appendix A

# THE MATHEMATICS OF COUNTERFORCE ATTACKS

Appendix A describes the mathematics behind estimating the potential benefits of counterforce attacks (both prelaunch attacks against enemy theater ballistic missiles (TBMs) and their transporter-erector-launchers (TELs), and postlaunch attacks against just the TELs).[1] The discussion will take place in three sections. The first will develop the formulas for calculating the expected number of TBMs that could be launched as a function of (1) the likely success of individual counterforce attacks ($P_k$), (2) the number of TELs in a TEL/TBM cluster, and (3) the number of available TBMs in the same cluster. These formulas usually cannot be stated in closed form and require solution by computer. Because of their nature, this limits the magnitude of the numbers of TELs and TBMs that can be calculated with these formulas.

The second section will describe a Markov matrix approach to solving the same problem. This approach also requires computer solution, but alleviates some of the constraints on the numbers of TELs and TBMs that can be included in the calculation.

The third section will describe a Monte Carlo computer model that also solves this problem. This approach has the benefit of relatively rapid calculation of outcomes for large numbers of TBMs and TELs, with the constraint that the outcomes are inherently random. Outputs from a number of runs will be described.

[1]Common usage is to call prelaunch attacks *counterforce*, and postlaunch attacks *counterbattery*. We will generally follow this usage.

## SECTION ONE: SOLVING THE PROBLEM BY ALGEBRAIC METHODS

### Part One: The Mathematics of Postlaunch Attacks Against the TELs Only

**A. Assumptions.** We start with the following assumptions.

1. The TBMs and TELs are organizationally formed into *clusters*. A cluster is a unit of TBMs and TELs that operates autonomously. For purposes of reloading, every TEL in the cluster is assumed to have access to every TBM in the cluster. Once all TBMs in the cluster are used, the cluster is empty (i.e., it is not resupplied). Also, if all TELs belonging to the cluster are killed, any remaining missiles in the cluster are lost.
2. The number of TELs in the cluster is T.
3. The number of TBMs in the cluster is N. N is larger than T (reloads are assumed to exist).
4. Each TEL is loaded with a TBM when it leaves its hide/resupply site. The TELs are not subjected to attrition on their way to the launch site and are assumed to successfully launch the TBM at the site. Counterbattery attacks are assumed to occur while the TEL is returning to its temporary hide or reload/resupply site.
5. All TELs are treated the same. Each survives the attempted counterbattery attack (after every launch) with probability $(1 - P_k)$.

These assumptions reflect the operational concept for counter-battery attacks—that the approximate location of the TEL is determined by backtracking the TBM while it is in powered flight (i.e., while it has a very bright, readily detectable signal that would be difficult to conceal or mask), and an airborne platform suitably armed flies to that location, tries to detect and identify the (now empty) TEL, and prosecutes the attack. The mathematics are generally applicable to alternative concepts of operations, but we will use this one for ease of description.

**B. The Mathematics for a Single TEL per Cluster.** Let P(K) = probability that exactly K TBMs are successfully launched. This can happen only if the TEL successfully launches its TBM, recovers and reloads (K – 1) times and is subsequently killed after the Kth launch. The only exception to this would be the case where the Kth launch is equal to N. Thus we address two separate cases, one where K < N and one where K = N.

For K < N:

$$P(K) = \left(1 - P_k\right)^{(K-1)} \cdot P_k \tag{A.1}$$

where

$$\left(1 - P_k\right)^{K-1}$$

is the probability that the TEL survives the first K – 1 recoveries from the launch site and $P_k$ is the probability that it gets killed on the last. Note that K runs from 1 to N, reflecting the fact that the first launch is not contested (all counterbattery attacks occur after the launch) and the TEL cannot launch more missiles than exist in the cluster.

For K = N:

$$P(K) = \left(1 - P_k\right)^{(N-1)} \tag{A.2}$$

In other words, to launch all N TBMs, the TEL needs only to survive the first (N – 1) counterbattery attacks. Whether it survives the counterbattery attack after the last launch makes no difference to the number of TBMs launched (i.e., the cluster is empty).

**Expected value for 1 TEL and N TBMs:** Let $E_1(K)$ = expected number of TBMs that are successfully launched. $E_1(K)$ is given by the following sum:

$$E_1(K) = \sum_{K=1}^{N} K \cdot P(K) = \sum_{K=1}^{N-1} K \cdot P_k \cdot \left(1 - P_k\right)^{K-1} + N \cdot \left(1 - P_k\right)^{N-1} \tag{A.3}$$

where Eqs (A.1) and (A.2) have been substituted into the first summation. To solve this equation in closed form, recall the following series expansions:

$$\frac{1}{(1-x)^2} = 1 + 2x + 3x^2 + 4x^3 + \ldots = \sum_{i=1}^{\infty} i x^{i-1} \tag{A.4}$$

$$\frac{1}{1-x} = 1 + x + x^2 + x^3 + \ldots = \sum_{i=1}^{\infty} x^{i-1} \tag{A.5}$$

Also note the following identity:

$$\sum_{i=1}^{N-1} f(i) = \sum_{i=1}^{\infty} f(i) - \sum_{i=N}^{\infty} f(i) \tag{A.6}$$

Using Eqs (A.4), (A.5), and (A.6),

$$\sum_{i=1}^{N-1} i \cdot x \cdot (1-x)^{i-1} = x \sum_{i=1}^{\infty} i \cdot (1-x)^{i-1} - x \sum_{i=N}^{\infty} i \cdot (1-x)^{i-1} \tag{A.7}$$

To solve these two new summations, note that the first summation is simply the expansion shown in Eq. (A.4). The second summation has a similar character, but the series does not start at 1. To solve it, change variables in the last summation, letting j = i + 1 − N. This yields

$$\begin{aligned} &= \frac{x}{(1-(1-x))^2} - x \sum_{j=1}^{\infty} (N-1+j)(1-x)^{N-1+j-1} \\ &= \frac{1}{x} - x \cdot (1-x)^{N-1} \sum_{j=1}^{\infty} (N-1+j)(1-x)^{j-1} \end{aligned} \tag{A.8}$$

Expanding the remaining sum into two new sums, we get

$$\begin{aligned} &= \frac{1}{x} - x \cdot (1-x)^{N-1} \left\{ \sum_{j=1}^{\infty} (N-1)(1-x)^{j-1} + \sum_{j=1}^{\infty} j(1-x)^{j-1} \right\} \\ &= \frac{1}{x} - x \cdot (1-x)^{N-1} \left\{ \frac{N-1}{x} + \frac{1}{x^2} \right\} \end{aligned} \qquad (A.9)$$

where we have used Eqs. (A.5) and (A.4) on the two sums, respectively. If we now substitute the pertinent variables, that is, $P_k = x$ and $N = n$, into equation (A.9), and substitute this into Eq. (A.3), we obtain

$$E_1(K) = N \cdot (1-P_k)^{N-1} + \frac{1}{P_k} - (N-1)(1-P_k)^{N-1} - \frac{(1-P_k)^{N-1}}{P_k} \qquad (A.10)$$

which simplifies to our final form:

$$E_1(K) = \frac{1-(1-P_k)^N}{P_k} \qquad (A.11)$$

Note that if N is large, the second term in the numerator becomes nearly zero, yielding the result:

$$\lim_{N \to \infty} E_1(K) = \frac{1}{P_k} \qquad (A.12)$$

This simple expression for the expectation can easily be derived from probability theory. For our purposes, however, Eq. (A.11) is of more interest, because we generally believe that TBM stockpiles will be finite and not out of proportion to the number of available TELs.

In addition to the expected value, it is useful to calculate the variance (or standard deviation). If Var(K) is the variance of these outcomes, then

$$\mathrm{Var}(K) = \sum_{K=1}^{N} K^2 P(K) - \left(E(K)\right)^2 \tag{A.13}$$

After substituting Eqs. (A.1), (A.2), and (A.10) into Eq. (A.13), we find that no simple closed-form solution exists. Thus,

$$\mathrm{Var}(K) = N^2\left(1 - P_k\right)^{N-1} + P_k \sum_{K=1}^{N-1} K^2\left(1 - P_k\right)^{K-1} - \left(\frac{1-\left(1-P_k\right)^N}{P_k}\right)^2 \tag{A.14}$$

Note that

$$\lim_{P_k \to 0} \mathrm{Var}(K) = 0$$

and

$$\lim_{P_k \to 1} \mathrm{Var}(K) = 0,$$

which is as it should be.

**C. The Mathematics for 2 TELs per Cluster.** The mathematics involving multiple TELs per cluster adds the complication that there are multiple ways to obtain exactly K successful launches. As above, let P(K) equal the probability that exactly K TBM launches are made. Then

For $2 \leq K < N$:

$$P(K) = \sum_{i=1}^{K-1} \left\{\mathrm{Prob}\left(\text{TEL \# 1 launches exactly i TBMs}\right)\right\} \cdot \left\{\mathrm{Prob}\left(\text{TEL \# 2 launches K} - \text{i TBMs}\right)\right\} \tag{A.15}$$

which, from above, we know leads to

$$P(K) = \sum_{i=1}^{K-1} \left\{ \left(1 - P_k\right)^{i-1} P_k \right\} \cdot \left\{ \left(1 - P_k\right)^{K-1-i} P_k \right\}$$

$$= \sum_{i=1}^{K-1} \left(1 - P_k\right)^{K-2} P_k^{\,2} = \left(K - 1\right) P_k^{\,2} \left(1 - P_k\right)^{K-2} \tag{A.16}$$

Note that K starts at 2, reflecting the assumption that both TELs launch their first TBMs without being contested.

For K = N:

$$P(N) = 1 - \sum_{K=2}^{N-1} P(K) \tag{A.17}$$

which, after substituting Eq. (A.16) into (A.17), becomes

$$P(N) = 1 - \sum_{K=2}^{N-1} (K - 1) P_k^{\,2} \left(1 - P_k\right)^{K-2} \tag{A.18}$$

which, after a suitable change of variables $(j = K - 1)$, simplifies to

$$P(N) = 1 - P_k^{\,2} \sum_{j=1}^{N-2} j \left(1 - P_k\right)^{j-1} \tag{A.19}$$

Before proceeding to further refine Eq. (A.19), let us back up to Eqs. (A.16) and (A.18) and write the expression for the expected number of TBMs launched. As before,

$$E_2(K) = \sum_{K=2}^{N} K \cdot P(K) = N - N P_k^{\,2} \sum_{K=2}^{N-1} \left(K - 1\right)\left(1 - P_k\right)^{K-2}$$

$$+ P_k^{\,2} \sum_{K=2}^{N-1} K\left(K - 1\right)\left(1 - P_k\right)^{K-2} \tag{A.20}$$

which reduces to

$$E_2(K) = N - P_k^{\,2} \sum_{K=2}^{N-1} (N-K)(K-1)(1-P_k)^{K-2} \tag{A.21}$$

This form for the expected value will reappear later when the summation terms cannot be readily simplified.

Equation (A.19) is similar to Eq. (A.7), and can be simplified in the same manner. Expanding into two summations and changing variables in the second (to get the summation from one to infinity), we obtain

$$\sum_{j=1}^{N-2} j(1-P_k)^{j-1} = \sum_{j=1}^{\infty} j(1-P_k)^{j-1} - \sum_{j=1}^{\infty} (j+N-2)(1-P_k)^{N-2+j-1} \tag{A.22}$$

The first term in Eq. (A.22) is simply

$$\frac{1}{P_k^{\,2}}.$$

To solve the second summation, separate it into two new summations as follows:

$$\sum_{j=1}^{\infty} (j+N-2)(1-P_k)^{N-2+j-1} = (1-P_k)^{N-2} \left\{ \sum_{j=1}^{\infty} j(1-P_k)^{j-1} + \sum_{j=1}^{\infty} (N-2)(1-P_k)^{j-1} \right\} \tag{A.23}$$

$$= (1-P_k)^{N-2} \left\{ \frac{1}{P_k^{\,2}} + \frac{(N-2)}{P_k} \right\} \tag{A.24}$$

Substituting Eq. (A.24) back into Eq. (A.22) and that sum back into Eq. (A.19), we obtain

$$P(N) = \left(1 - P_k\right)^{N-2}\left\{1 + P_k\left(N - 2\right)\right\} \tag{A.25}$$

Thus, by suitable algebraic manipulation we have managed to obtain closed-form solutions for the probabilities of exactly K TBMs being launched for all values of K.

**Expected value for 2 TELs and N TBMs:** Substituting Eq. (A.25) into the formula for the expected value of the number of TBMs launched yields

$$\begin{aligned} E_2(K) = \sum_{K=2}^{N} K \cdot P(K) = N\left(1 - P_k\right)^{N-2}\left\{1 + P_k\left(N - 2\right)\right\} \\ + \sum_{K=2}^{N-1} K\left(K - 1\right)P_k^{\,2}\left(1 - P_k\right)^{K-2} \end{aligned} \tag{A.26}$$

Alas, the authors are not aware of a closed-form solution for this summation. However the equation is easily calculated by numerical means, and the authors have built a simple BASIC program for this purpose (see Appendix B). Also note that Eq. (A.26) is not much simpler to calculate than Eq. (A.21).

## D. The Mathematics for 3 TELs per Cluster.

For $3 \le K < N$: The above approach of deriving the probability of exactly K launches will continue to be followed, but with 3 TELs it is necessary to introduce another variable. Thus, let $I_1$, $I_2$, and $I_3$ be the number of TBMs launched by TELs 1, 2, and 3, respectively. Obviously, for any K,

$$I_1 + I_2 + I_3 = K \tag{A.27}$$

Thus,

$$P(K) = \sum_{I_1=1}^{K-2} \Big\{ \text{Prob}\big(\text{TEL \# 1 launchs exactly } I_1 \text{ TBMs}\big) \cdot$$

$$\sum_{I_2=1}^{K-1-I_1} \Big[ \text{Prob}\big(\text{TEL \# 2 launches exactly } I_2 \text{ TBMs}\big) \cdot$$

$$\text{Prob}\big(\text{TEL \# 3 launches } I_3 \text{ TBMs}\big)\Big]\Big\}$$

And recognizing as before that

$$\text{Prob(TEL \#1 launches exactly } I_1 \text{ TBMs)} = P_k\left(1 - P_k\right)^{I_1 - 1}$$

$$\text{Prob(TEL \#2 launches exactly } I_2 \text{ TBMs)} = P_k\left(1 - P_k\right)^{I_2 - 1}$$

$$\text{Prob(TEL \#3 launches exactly } I_3 \text{ TBMs)} = P_k\left(1 - P_k\right)^{K - I_1 - I_2 - 1}$$

then

$$P(K) = \sum_{I_1=1}^{K-2} P_k\left(1 - P_k\right)^{I_1 - 1} \sum_{I_2=1}^{K-1-I_1} P_k\left(1 - P_k\right)^{I_2 - 1} P_k\left(1 - P_k\right)^{K - I_1 - I_2 - 1} \tag{A.28}$$

Pulling out the appropriate terms from the summations, we obtain

$$P(K) = P_k^{\,3}\left(1 - P_k\right)^{K-3} \sum_{I_1=1}^{K-2} \sum_{I_2=1}^{K-I_1-1} = P_k^{\,3}\left(1 - P_k\right)^{K-3} \sum_{I_1=1}^{K-2}\left(K - 1 - I_1\right) \tag{A.29}$$

Expanding the summation yields

$$P(K) = P_k^{\,3}\left(1 - P_k\right)^{K-3} \frac{(K-1)(K-2)}{2} \qquad \text{(A.30)}^2$$

---

[2]In calculating the sum or a series of consecutive integers, it is useful to recall the "trick" attributed to Gauss when he was in the first grade. Add the sum to itself with

For K = N:

By definition, and using Eq. (A.30) for P(K) where K < N,

$$P(N) = 1 - \sum_{K=3}^{N-1} P(K) \tag{A.31}$$

$$= 1 - \sum_{K=3}^{N} P_k^{\ 3}\left(1 - P_k\right)^{K-3} \frac{(K-1)(K-2)}{2}$$

**Expected Value for 3 TELs and N TBMs:** Substituting into the expression for the expected value of number of TBMs launched,

$$E_3(K) = \sum_{K=3}^{N} K \cdot P(K)$$

$$= N - N \cdot \sum_{K=3}^{N-1} P_k^{\ 3}\left(1 - P_k\right)^{K-3} \frac{(K-1)(K-2)}{2}$$

$$+ \sum_{K=3}^{N-1} K \cdot P_k^{\ 3}\left(1 - P_k\right)^{K-3} \frac{(K-1)(K-2)}{2} \tag{A.32}$$

$$= N - P_k^{\ 3} \sum_{K=3}^{N} (N-K) \frac{(K-1)(K-2)}{2}\left(1 - P_k\right)^{K-3}$$

The authors do not know of a way to solve this series in a simpler form. However, computer codes have been built to evaluate it (see Apendix B).

---

the series reversed; that is, the first number is added to the last, the second to the next to last, etc. The resulting sums are all constant, and there are exactly n of them, where n is the number of integers in the series. Therefore, the new sum is just the product of the number of integers times the sum of the first and last term in the series. To get the answer, simply divide by two, reflecting the fact that you have added the series twice.

**E. The Mathematics of 4 TELs per Complex.** The approach taken is identical to that in D. above—solve first for the probability of exactly K launch successes (K less than or equal to the total number of TBMs in the complex), and then plug the solution into the expected-value equation. Thus,

For $4 \le K < N$:

$$P(K) = \sum_{I_1=1}^{K-3} \Big\{ \text{Prob}\big(\text{TEL \# 1 launches exactly } I_1 \text{ TBMs}\big) \cdot \sum_{I_2=1}^{K-1-I_1} \text{Prob}\big(\text{TEL \# 2 launches exactly } I_2 \text{ TBMs}\big) \cdot \sum_{I_3=1}^{K-1-I_1-I_2} \Big[ \text{Prob}\big(\text{TEL \# 3 launches exactly } I_3 \text{ TBMs}\big) \cdot \text{Prob}\big(\text{TEL \# 4 launches exactly } I_4 \text{ TBMs}\big) \Big] \Big\}$$

$$= \sum_{I_1=1}^{K-3} P_k (1-P_k)^{I_1-1} \sum_{I_2=1}^{K-2-I_1} P_k (1-P_k)^{I_2-1} \sum_{I_3=1}^{K-1-I_1-I_2} P_k (1-P_k)^{I_3-1} P_k (1-P_k)^{K-I_1-I_2-I_3-1}$$

$$= P_k^{\,4} (1-P_k)^{K-4} \sum_{I_1=1}^{K-3} \sum_{I_2=1}^{K-2-I_1} \sum_{I_3=1}^{K-1-I_1-I_2}$$

$$= P_k^{\,4} (1-P_k)^{K-4} \sum_{I_1=1}^{K-3} \sum_{I_2=1}^{K-2-I_1} (K-1-I_1-I_2)$$

which yields

$$P(K) = P_k^{\,4}\left(1 - P_k\right)^{K-4} \sum_{I_1=1}^{K-3} \frac{\left(K - 1 - I_1\right)\left(K - 2 - I_2\right)}{2} \tag{A.33}$$

For K = N:

$$\begin{aligned} P(N) &= 1 - \sum_{K=4}^{N-1} P(K) \\ &= 1 - \sum_{K=4}^{N-1} P_k^{\,4}\left(1 - P_k\right)^{K-4} \sum_{I_1=1}^{K-3} \frac{\left(K - 1 - I_1\right)\left(K - 2 - I_1\right)}{2} \end{aligned} \tag{A.34}$$

**Expected value for 4 TELs and N TBMs:**

As above,

$$\begin{aligned} E_4(K) = N - \sum_{K=4}^{N-1} \left(N - K\right) \cdot P_k^{\,4}\left(1 - P_k\right)^{K-4} \\ \sum_{I_1=1}^{K-3} \frac{\left(K - 1 - I_1\right)\left(K - 2 - I_1\right)}{2} \end{aligned} \tag{A.35}$$

There is no way to significantly simplify this expression. It is, however, readily amenable to numerical calculation.

**F. Generalization of Mathematics to Larger Numbers of TELs per Cluster.** The generalization should be obvious. The reader can confirm the following equations:

For K < N:

$$P(K) = P_k^{\ T}\left(1-P_k\right)^{K-T} \sum_{I_1}^{[K-(T-1)]} \sum_{I_2}^{[K-(T-2)-I_1]} \sum_{I_3}^{[K-(T-3)-I_1-I_2]} \cdots \sum_{I_{T-1}}^{[K-1-I_1-\cdots-I_{T-2}]} \quad (A.36)$$

For K = N:

$$P(N) = 1 - \sum_{K=T}^{N} P_k^{\ T}\left(1-P_k\right)^{K-T} \sum_{I_1}^{[K-(T-1)]} \sum_{I_2}^{[K-(T-2)-I_1]} \sum_{I_3}^{[K-(T-3)-I_1-I_2]} \cdots \sum_{I_{T-1}}^{[K-1-I_1-\cdots-I_{T-2}]} \quad (A.37)$$

If, for each notation, we define the summations as $S_b(K,T)$, where the subscript b denotes that this sum applies for counterbattery calculations, then

$$S_b(K,T) = \sum_{I_1}^{[K-(T-1)]} \sum_{I_2}^{[K-(T-2)-I_1]} \sum_{I_3}^{[K-(T-3)-I_1-I_2]} \cdots \sum_{I_{T-1}}^{[K-1-I_1-\cdots-I_{T-2}]} \quad (A.38)$$

and

$$E_T(K) = N - \sum_{K=T}^{N-1} (N-K) P_k^{\ T}\left(1-P_k\right)^{(K-T)} S_b(K,T) \quad (A.39)$$

Equation (A.39) is the general solution to this expected-value problem. The equation is amenable to computer calculation, although the time required for its calculation grows substantially as T and N increase. For practical purposes, calculations for T > 10 are probably not best done by this equation.

## Part Two: Extending the Mathematics for Prelaunch Attacks Against Both the TBMs and TELs

**A. Assumptions.** The first three assumptions in Part One are the same. However, because the attack can kill both the TELs and the TBM, changes to assumptions 4 and 5 are needed.

4. Each TEL possesses a TBM. The TELs are subjected to attrition on their way to the launch site. A successful prelaunch counterforce attack will kill both the TBM and the TEL.
5. All TBM/TELs are treated the same. Each survives the attempted counterforce attack (before each launch) with probability $(1 - P_k)$.

**B. The Mathematics for a Single TEL per Cluster.** Much of the formalism is identical to that used above. The primary changes concern the range for the summations (they start with zero instead of one, as it is possible that no TBMs will be launched). We again start by looking at the probability that exactly K TBMs are launched.

For $0 \le K < N$:

$$P(K) = \left(1 - P_k\right)^K P_k \qquad \text{(A.40)}$$

where the TEL successfully survives the first K sorties out of its hide site, launching K TBMs, but is discovered and killed on the K + 1 sortie.

For $K = N$:

$$P(N) = \left(1 - P_k\right)^N \qquad \text{(A.41)}$$

where the TEL survives all N sorties, launching all N TBMs.

**Expected-value equation:**

$$E_1(K) = \sum_{K=0}^{N} K \cdot P(K) = N \cdot \left(1 - P_k\right)^N + \sum_{K=0}^{N-1} K \cdot \left(1 - P_k\right)^K P_k \qquad \text{(A.42)}$$

The summation at the end of Eq. (A.42) can be solved in closed form by the techniques used above[3]). Solving this equation yields the following result:

$$E_1(K) = \left(\frac{1}{P_k} - 1\right) \cdot \left(1 - \left(1 - P_k\right)^N\right) \tag{A.43}$$

Comparing this with Eq. (A.11) shows the addition of a –1 term to

$$\frac{1}{P_k},$$

reflecting the fact that killing the TEL results in loss of one TBM as well.

**C. The Mathematics for 2 TELs.** Much of the above applies for 2 TELs as well. However, there is at least one important difference. Because a TBM is killed every time that a TEL is killed, it is important to consider separately the case where exactly 1 (out of 2) TELs survives to TBM exhaustion.

As before, for $0 \le K < (N - 1)$

$$P(K) = \sum_{i=0}^{K} \left(1 - P_k\right)^i P_k \left(1 - P_k\right)^{K-i} P_k$$

$$= \left(K + 1\right) P_k^{\,2} \left(1 - P_k\right)^K \tag{A.44}$$

For K = N – 1, one of the TELs must be killed, and one must have survived to TBM exhaustion. Thus,

$$P(N-1) = \sum_{i=0}^{N-1} \left(1 - P_k\right)^i P_k \left(1 - P_k\right)^{K-i}$$

---

[3]Extract a $\left(1 - P_k\right)$ from the summation.

where the kill of the first TEL can occur anytime, including up to the time when the TEL is carrying the last TBM to the launch stand. Thus,

$$P(N-1) = NP_k\left(1-P_k\right)^{N-1} \tag{A.45}$$

Finally, for all TBMs to be launched, all TELs must survive. Thus,

$$P(N) = \left(1-P_k\right)^N \tag{A.46}$$

Using Eqs. (A.44), (A.45), and (A.46), the expected value can be written as

$$\begin{aligned} E_2(K) = {} & N\left(1-P_k\right)^N + N\left(N-1\right)P_k\left(1-P_k\right)^{N-1} \\ & + \sum_{K=0}^{N-2} K\left(K+1\right)P_k^{\,2}\left(1-P_k\right)^K \end{aligned} \tag{A.47}$$

This equation cannot be significantly simplified.

**D. The Mathematics for 3 TELs.** The equations for 2 TELs point the way, but now we must consider as separate cases the situation where exactly 1 and 2 TELs survive to TBM exhaustion.

If all TELs are killed before TBM exhaustion, then

For $K < N-2$:

$$\begin{aligned} P(K) &= \sum_{i=0}^{K} \left(1-P_k\right)^i P_k \sum_{j=0}^{K-i} \left(1-P_k\right)^j P_k \left(1-P_k\right)^{K-i-j} P_k \\ &= \left(1-P_k\right)^K P_k^{\,3} \frac{\left(K+1\right)\left(K+2\right)}{2} \end{aligned} \tag{A.48}$$

For $K = N-2$ (2 TELs killed, 1 survives to TBM exhaustion):

$$P(N-2) = \sum_{i=0}^{N-2} \left(1-P_k\right)^i P_k \sum_{j=0}^{N-2-i} \left(1-P_k\right)^j P_k \left(1-P_k\right)^{N-2-i-j}$$

$$= \left(1-P_k\right)^{N-2} P_k^{\,2} \frac{N\left(N-1\right)}{2} \qquad \text{(A.49)}$$

For K = N – 1 (1 TEL killed, the remaining 2 survive to TBM exhaustion), and for K = N (all TELs survive to TBM exhaustion), the equations are identical to those for the 2-TEL case above; that is,

$$P(N-1) = NP_k\left(1-P_k\right)^{N-1} \qquad \text{(A.50)}$$

and

$$P(N) = \left(1-P_k\right)^N \qquad \text{(A.51)}$$

The expected value is

$$E_3(K) = N\left(1-P_k\right)^N + N\left(N-1\right)\left(1-P_k\right)^{N-1} P_k$$

$$+ \frac{N\left(N-1\right)\left(N-2\right)}{2}\left(1-P_k\right)^{N-2} P_k^{\,2}$$

$$+ \sum_{K=1}^{N-3} \frac{K\left(K+1\right)\left(K+2\right)}{2}\left(1-P_k\right)^{K-3} P_k^{\,3} \qquad \text{(A.52)}$$

**E. The Mathematics for N TELs.** The symmetry in the above equations should now be evident. Using the notation from the first part of this section, we can now write a general expression for the various probabilities and the expected value.

For $0 \le K \le N - T$:

$$P(K) = \left(1 - P_k\right)^K P_k{}^T \sum_{I_1=0}^{K} \sum_{I_2=0}^{K-I_1} \cdots \sum_{I_{T-2}}^{K-I_1-I_2-\cdots-I_{T-3}} \left(K + 1 - I_1 - I_2 - \cdots - I_{T-2}\right) \quad \text{(A.53)}$$

Define a new variable for the summation; i.e.,

$$S_f(K, T) = \sum_{I_1=0}^{K} \sum_{I_2=0}^{K-I_1} \cdots \sum_{I_{T-2}}^{K-I_1-I_2-\cdots-I_{T-3}} \left(K + 1 - I_1 - I_2 - \cdots - I_{T-2}\right) \quad \text{(A.54)}$$

then

$$P(K) = P_k{}^T \left(1 - P_k\right)^K S_f(K, T) \quad \text{(A.55)}$$

For K = N – (T – 1), where all but one TEL is killed before TBM exhaustion,

$$P(K) = P_k{}^{T-1} \left(1 - P_k\right)^K S_f(K, T) \quad \text{(A.56)}$$

For K = N – (T – 2),

$$P(N - (T - 2)) = P_k{}^{T-2} \left(1 - P_k\right)^{N-(T-2)} S_f(K, T - 1) \quad \text{(A.57)}$$

And, in general, for $N - T < K < N$,

$$P(N - (T - j)) = P_k{}^{T-j} \left(1 - P_k\right)^{N-(T-j)} S_f(K, T - j + 1) \quad \text{(A.58)}$$

where $1 \le j \le T - 1$

For $K = N$,

$$P(N) = \left(1 - P_k\right)^N \quad \text{(A.59)}$$

## Part Three: The Mathematics of Combined Prelaunch and Postlaunch Counterforce Attacks

Because there are simpler ways than the algebraic approach to solve the combined prelaunch counterforce and postlaunch counterbattery attacks, we will consider only the single TEL case.

Let $P_1$ be the probability of successful prelaunch counterforce attack against the TEL, and $P_2$ the probability of successful counterbattery attack against the TEL after it has launched the TBM. Let $Q_1 = \left(1 - P_1\right)$ and $Q_2 = \left(1 - P_2\right)$. In addition, define $Q_3 = Q_1 Q_2$. Then, for $K < N$,

$$P(K) = \left(1 - P_1\right)^K \left(1 - P_2\right)^{K-1} \left\{P_2 + \left(1 - P_2\right)P_1\right\} \quad \text{(A.60)}$$

where the only way that exactly K TBMs can be launched is for the TEL to survive K prelaunch counterforce engagements and K – 1 postlaunch engagements, only to be killed (the terms in the brackets) after the kth TBM launch by a counterbattery attack or by a prelaunch engagement on the TEL's way to launch its K + 1 TBM. This equation can be rewritten into the following form:

$$P(K) = Q_1\left(Q_1 Q_2\right)^{K-1}\left[1 - Q_1 Q_2\right]$$

or

$$P(K) = Q_1 {Q_3}^{K-1}\left(1 - Q_3\right) \quad \text{(A.61)}$$

For $K = N$, the formula is simply

$$P(N) = Q_1 {Q_3}^{N-1} \quad \text{(A.62)}$$

As above, we can derive the expected number of TBMs launched by suitable algebra, resulting in the following equation:

$$E_1(K) = \frac{Q_1\left(1 - Q_3{}^N\right)}{\left(1 - Q_3\right)} \tag{A.63}$$

A quick comparison between this equation and those for counterbattery and counterforce shows that this formula is identical to them if either kill probability is set to zero.

## Section Two: The Markov Matrix Approach for Obtaining Outcomes for Counterforce and Counterbattery Attacks[4]

A different, and in many ways more satisfying, approach to solving the problem is through the approach which might roughly be equated to a Markov matrix model. To describe this approach, we make the following assumptions.

1. At the start T TELs and N TBMs are assumed to exist and to be colocated at a common hide site (for analytic ease, we will assume that the so-called operational hide site and the resupply site are the same). We assume that $N > T$, that is there is more than one TBM per TEL.
2. Each step in the process starts by assuming that a single TEL, outfitted with a TBM, leaves its hide site, committed to launching the TBM. On route to the launch site, the TEL and its TBM are subjected to a counterforce attack. If the attack is successful, both the TEL and the TBM are killed. If the attack fails, we assume that the TEL successfully reaches the launch site and launches the TBM. After the TBM is launched, the TEL attampts to return to its hide site. We assume that the TEL is again subject to attack, and successfully returns to its hide only if it survives this attack as well. If the TEL reaches its hide site successfully, it is available to be reloaded with a new TBM and reused.

---

[4]The author owes a debt of gratitude to David Vaughan for suggesting this approach.

3. The TEL is not attacked except on its way to and from the launch site.
4. If an individual TEL is killed, a replacement TEL (if the initial stockpile of T TELs has not been exhausted) repeats the above process.
5. The process terminates when all T TELs have been killed or all the TBMs launched.

## Part One: The Postlaunch Counterbattery Attack Case Only

For ease of presentation, we start with the postlaunch-counterbattery-attack-only case. Let t(i,x) be the probability of exactly i TELs surviving after x attempts to launch a TBM. Assume for simplicity that T = 3, and let $P_s = (1 - P_k)$. Then

$$\begin{bmatrix} t(3,x) \\ t(2,x) \\ t(1,x) \\ t(0,x) \end{bmatrix} = \begin{bmatrix} P_s & 0 & 0 & 0 \\ P_k & P_s & 0 & 0 \\ 0 & P_k & P_s & 0 \\ 0 & 0 & P_k & 1 \end{bmatrix} \times \begin{bmatrix} t(3,x-1) \\ t(2,x-1) \\ t(1,x-1) \\ t(0,x-1) \end{bmatrix} \tag{A.64}$$

with the following initial conditions:

$$\begin{bmatrix} t(3,0) \\ t(2,0) \\ t(1,0) \\ t(0,0) \end{bmatrix} = \begin{bmatrix} 1 \\ 0 \\ 0 \\ 0 \end{bmatrix} \tag{A.65}$$

Define E(x) as the expected number of TBMs launched, given x attempts. Thus,

$$E(x) = \sum_{i=1}^{T} t(i, x-1) + E(x-1) \tag{A.66}$$

Alternatively, E(x) can be written as

$$E(x) = \big(1 - t(0, x - 1)\big) + E(x - 1) \tag{A.67}$$

More generally, assume that A(i,j) is the probability matrix that transforms t(j,x – 1) to t(i,x). Then

$$\vec{t}(i, x) = A(i, j) \cdot \vec{t}(j, x - 1) \tag{A.68}$$

Because the matrix A is relatively sparse, it is computationally simpler to pose the problem in the form of the following set of difference equations:

For $i = T$

$$t(T, x) = P_s t(T, x - 1) \tag{A.69}$$

For $0 < i < T$

$$t(i, x) = P_s t(i, x - 1) + P_k t(i + 1, x) \tag{A.70}$$

For $i = 0$

$$t(0, x) = P_k t(1, x - 1) + t(0, x - 1) \tag{A.71}$$

Finally, x runs from 1 to N.

## Part Two: Extending the Markov Matrix Model to Include Prelaunch Counterforce Attacks

The above approach is readily extended to include prelaunch counterforce attacks. Assuming that prelaunch and postlaunch attacks are successful with probabilities ${}^1P_k$ and ${}^2P_k$, respectively, then

For $i = T$

$$t(T, x) = \left(1 - {}^1P_k\right)\left(1 - {}^2P_k\right) t(T, x - 1) \tag{A.72}$$

For $0 < i < T$

$$t(i, x) = \left(1 - {}^1P_k\right)\left(1 - {}^2P_k\right) t(i, x - 1) + \left\{ {}^1P_k + \left(1 - {}^1P_k\right) {}^2P_k \right\} t(i + 1, x - 1) \tag{A.73}$$

For i = 0 $\quad t(0, x) = t(0, x-1) + \left\{ {}^1P_k + \left(1 - {}^1P_k\right)^2 P_k \right\} t(1, x-1)$ (A.74)

The coefficients of t(i,x) are the elements of the Markov matrix. The bracketed term can be rewritten into a somewhat simpler form; i.e.,

$$\left\{ {}^1P_k + \left(1 - {}^1P_k\right)^2 P_k \right\} = \left(1 - Q_1 Q_2\right) = \left(1 - Q_3\right)$$

where $Q_1$ and $Q_2$ are the TEL survival probabilities against prelaunch and postlaunch counterforce attacks, respectively, and $Q_3 = Q_1 Q_2$. Substituting $Q_1$, $Q_2$, and $Q_3$ into Eqs. (A.72), (A.73), and (A.74), we obtain the following compact form:

For i = T $\quad t(T, x) = Q_3 t(T, x-1)$ (A.75)

For 0 < i < T $\quad t(i, x) = Q_3 t(i, x-1) + \left(1 - Q_3\right) t(i+1, x-1)$ (A.76)

For i = 0 $\quad t(0, x) = t(0, x-1) + \left(1 - Q_3\right) t(1, x-1)$ (A.77)

The number of TBMs launched, E(x), takes a slightly different form:

$$E(x) = Q_1 \cdot \left[1 - t(0, x-1)\right] + E(x-1) \qquad \text{(A.78)}$$

where a TBM is successfully launched only if the TEL survives the pre-launch counterforce attack. Also note that the probability that exactly x TBMs are launched is just the probability that after x launch attempts the last TEL is killed. If we set f(x) as this probability, then

$$f(x) = t(0, x) - t(0, x-1) \qquad \text{(A.79)}$$

## Part Three: Solving the Markov Matrix

It is interesting to note that

$$\vec{t}(i, x) = A^x(i, j) \cdot \vec{t}(j, 0) \qquad \text{(A.80)}$$

where i and j are indices for the matrix roles and columns, respectively, (1,1 being at the top left part of the matrix) and where $A^x$ is just the matrix A multiplied by itself x times. The coefficients of $A^x$ have well-defined structures, easily obtained by simply carrying out the indicated multiplications. For those cases where the number of launch attempts, x, is larger than r, where r is the rank of the matrix A, the first r – 1 coefficients of the first column in $A^x$ are simply the first r – 1 terms in the expansion of

$$\left(Q_3 + P_3\right)^x,$$

respectively. The rth term is simply one minus the sum of the first r – 1 terms.[5] The coefficients of the second column are the same as the first, slid down one row, with the last coefficient equal to one minus the sum of the first r – 2 terms, and so forth for the remaining columns. Of course, the last column is all zeros except for the coefficient at the bottom.

Let $a_{i.j}(x)$ be the coefficients of the matrix $A^x$. Then assuming that x > r, the coefficients are as follows:

$$a_{i,j}(x) = \binom{x}{i-j} Q_3^{\,x-(i-j)} \left(1 - Q_3\right)^{i-j} \quad 1 \le i < r,\ 1 \le j \le i \qquad \text{(A.81)}$$

where $\binom{x}{i-j}$ is the binomial coefficient;

$$a_{i,j}(x) = 0 \qquad \text{for all } i < j < r; \qquad \text{(A.82)}$$

$$a_{r,j}(x) = 1 - \sum_{i=j}^{r-1} a_{i,j}(x) \qquad \text{for all } j < r; \qquad \text{(A.83)}$$

and

$$a_{r,r}(x) = 1 \qquad \text{(A.84)}$$

[5]The sum of all the coefficients in each column of the matrix must be equal to one.

The coefficients for $x < r$ are equally straightforward, but we will not offer them here. Figure A.1 shows the domains of the matrix associated with each of the above four equations.

The utility of the above solution is questionable, and calculations have simply used Eqs. (A.74) through (A.78).

## SECTION THREE: THE MONTE CARLO APPROACH TO SOLVING THIS PROBLEM

The third approach that can be taken is to simply write a Monte Carlo computer program where each TEL is subjected to prelaunch and postlaunch counterforce attacks. The flow chart of the program is shown in Figure A.2.

On each trial, the available TELs, loaded with TBMs, sortie from their hide sites to their launch sites. Random numbers are drawn for each

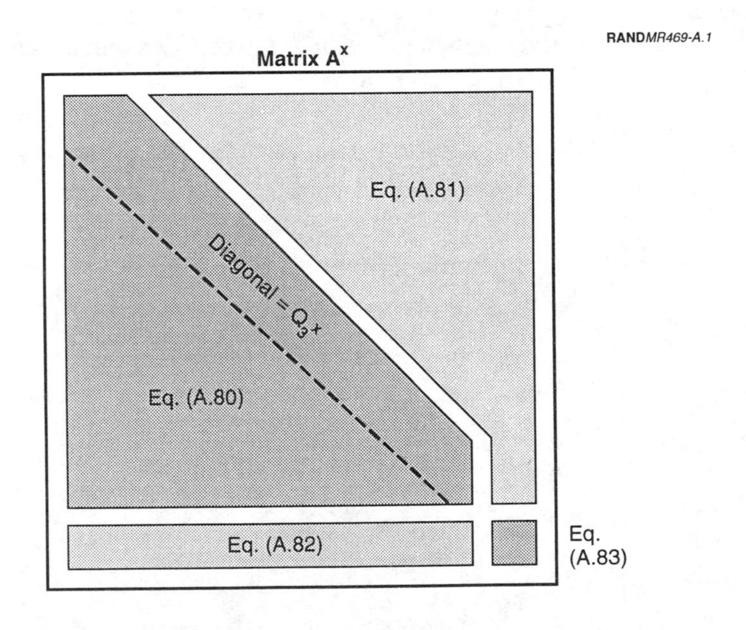


**Figure A.1—Domains of Matrix A Associated with Eqs. (A.80) Through (A.83)**

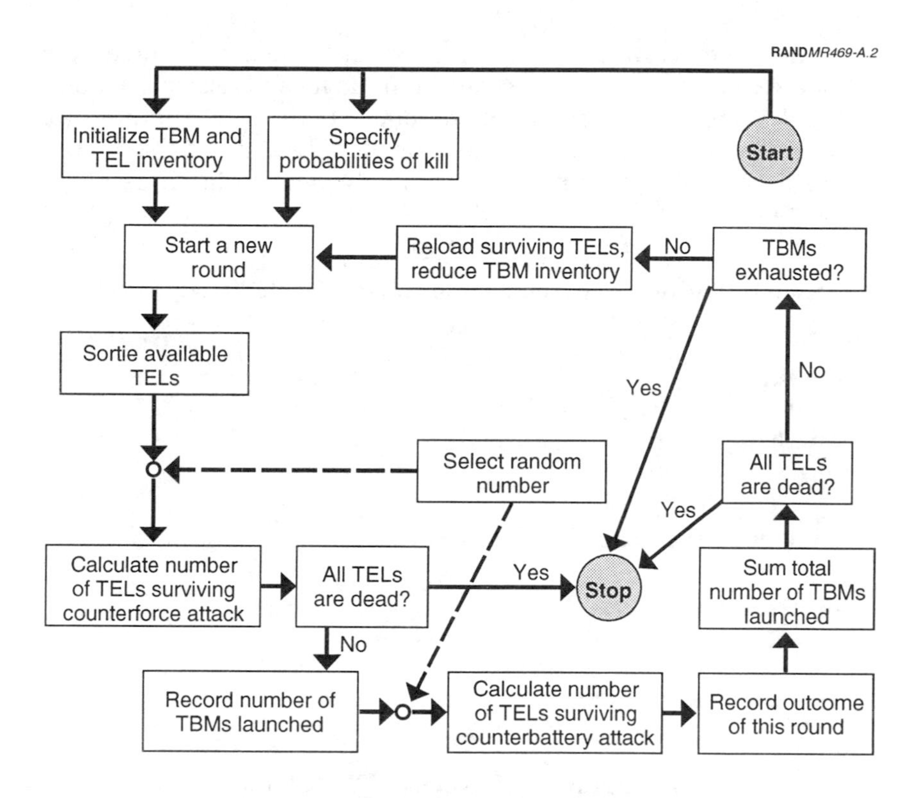


**Figure A.2—Flow Chart for Monte Carlo Calculation**

TEL to determine whether the prelaunch counterforce attack against that TEL succeeds. Assuming that the TEL survives, it launches its TBM unmolested. Subsequent to the launch, the TEL is subjected to a counterbattery attack. A new random number is drawn for each TEL, determining whether it survives that attack. Those TELs that survive return to their hide/resupply site, reload, and are available to participate in the next sortie. The numbers of TBMs launched are finished when one of three conditions is met: (1) all available TBMs are either launched or killed during the prelaunch attack phase, (2) all TELs are dead, or (3) the maximum number of trials (specified by the user) has been reached (not shown on Figure A.2).

Multiple trials are run, usually somewhere between 100 and 1000 separate runs, to obtain average outcomes not significantly different from the expected values. In addition to the statistical averages, the standard deviation of the outcomes from the mean is also calculated. Table A.1 shows an example of the data obtained by running the model. It shows that after 500 trials, the model estimated that on the average 49.57 TBMs would be launched, given the input assump-

**Table A.1**

**Sample Outcome from Monte Carlo Run for Counterbattery Attack**

**Input Data**

Total TELs in cluster = **100**
Total TBMs in cluster = **10**
Max. salvos = **50**
$P_k$ vs. TEL = **0.2**
Number of trials = **500**

**Output Data**

Average number of TBMs fired = **49.57**
Standard deviation = **13.241**
Standard deviation mean = **0.5922**

**Histogram of Number of TBMs Fired**

| Tens | Ones Count | | | | | | | | | |
|---|---|---|---|---|---|---|---|---|---|---|
| Count | **0** | **1** | **2** | **3** | **4** | **5** | **6** | **7** | **8** | **9** |
| **0** | 0 | 0 | 0 | 0 | 0 | 0 | 0 | 0 | 0 | 0 |
| **10** | 0 | 0 | 0 | 0 | 0 | 0 | 0 | 0 | 1 | 0 |
| **20** | 1 | 2 | 1 | 3 | 1 | 2 | 3 | 1 | 6 | 2 |
| **30** | 5 | 12 | 8 | 10 | 7 | 7 | 12 | 12 | 9 | 13 |
| **40** | 6 | 14 | 11 | 12 | 11 | 16 | 19 | 19 | 18 | 15 |
| **50** | 16 | 12 | 21 | 17 | 10 | 13 | 10 | 17 | 11 | 7 |
| **60** | 13 | 10 | 8 | 9 | 7 | 9 | 7 | 3 | 8 | 1 |
| **70** | 2 | 2 | 3 | 3 | 3 | 0 | 2 | 3 | 1 | 2 |
| **80** | 0 | 0 | 1 | 0 | 2 | 1 | 0 | 1 | 3 | 0 |
| **90** | 0 | 0 | 1 | 1 | 0 | 0 | 0 | 0 | 0 | 0 |
| **100** | 1 | 0 | 0 | 0 | 0 | 0 | 0 | 0 | 0 | 0 |

tions. This outcome is close to the value obtained by other analytic means. The estimated deviation of the results about this means is estimated to be 13.241 launches, a substantial number. Note that the model also estimates the standard deviation of the estimate of the mean—how far off our estimate is likely to be.

The histogram of the number of TBMs launched is also shown. The first column represents the number at the start of the row, and the numbers in the top row represent the number to be added to the first column. Thus, to find out how many times exactly 55 TBMs were launched, read down the first column to 50, then across the first row to 5. You should find that for these 500 trials exactly 13 resulted in producing 55 TBM launches. It is interesting to note that one trial produced 100 launches, a very unlikely outcome, and the minimum number of launches was 18, also a very unlikely outcome given that the first 10 launches occur before any counterbattery attacks occur.[6]

---

[6]For these input assumptions, there is less than one chance out of four that in 500 trials as few as 18 TBMs would be all that were launched.

Appendix B

# COMPUTER PROGRAMS FOR CALCULATING OUTCOMES FROM PRELAUNCH AND POSTLAUNCH COUNTERFORCE ATTACKS AGAINST TBMs AND TELs

**1. Simple counterforce/counterbattery model to calculate TBMs launched in the single TEL case.**

```
CLS
OPEN "clip:" FOR OUTPUT AS #1

'input pre-launch Pk, vary over counterbattery Pk
INPUT "Prob of pre-launch counterforce kill (P1)";p1
DIM E(20)

'start calculation at Pk less than one to avoid zero divisor
p2=.9

WHILE p2>0

q1=1-p1
q2=1-p2
q3=q1*q2

'Calculate for 1-10 TBMs per TEL
FOR N=1 TO 10
     E(N)=q1*(1-q3^N)/(1-q3)
     NEXT N
E(11)=1/(1-q3)

'Write to clipboard for later inclusion in EXCEL or CRICKET graphs
WRITE #1, p1,p2,E(1),E(2),E(3),E(4),E(5),E(6),E(7),E(8),E(9),E(10),E(11)
```

```
'Adjust step size to reflect sensitivities at small Pk
k=.1
IF p2<.3 THEN k=.02
IF p2<.08 THEN k=.01
p2=p2-k

WEND

CLOSE #1
```

**2. BASIC program to calculate expected number of TBMs launched, given a postlaunch counterbattery attack.**

This program does the following things:

1. It is based on mathematical equations for counterbattery attacks only.
2. Intermediate steps calculated the probabilities that specific outcomes are achieved. This information is not available in other calculations.
3. The program is limited to six TELs. Generalizations beyond six are straightforward, but the compute time would grow significantly.
4. Prints results to clipboard, to permit use of EXCEL or CRICKET graphics.
5. Automatically calculates outcomes for various Pks.

```
' Program to Calculate Expected Value Outcome of Counterbattery
    Attacks
CLS

OPEN "CLIP:" FOR OUTPUT AS #1
DIM z1(200)
DIM z2(200)
DIM z3(200)
DIM z4(200)
DIM z5(200)
DIM z6(200)
DIM z7(200)
DIM z8(200)
```

```
'Input variables
INPUT "Number of TBMs:", N
x=0
10 Pk=.9

'Start of calculation
WHILE Pk>0
x=x+1

'Calculation for 1 TEL
E1=(1-(1-Pk)^N)/Pk

'Calculation for 2 TELs
B2=0
FOR K=2 TO N-1
  B2=B2+(N-K)*(K-1)*(1-Pk)^(K-2)
  NEXT K
E2=N-B2*Pk^2

'Calculation for 3 TELs
B3=0
FOR K=3 TO N-1
  B3=B3+(N-K)*(K-1)*(K-2)*(1-Pk)^(K-3)
  NEXT K
E3=N-(B3*Pk^3)/2

'Calculation for 4 TELs
B4=0
FOR K=4 TO N-1
  C4=0
  FOR i=1 TO K-3
    C4=C4+(K-1-i)*(K-2-i)
    NEXT i
  B4=B4+(N-K)*C4*(1-Pk)^(K-4)
  NEXT K
E4=N-(B4*Pk^4)/2

'Calculation for 5 TELs
B5=0
FOR K=5 TO N-1
```

```
    C5=0
    FOR i=1 TO K-4
      D5=0
      FOR j=1 TO K-3-i
        D5=D5+(K-1-i-j)*(K-2-i-j)
        NEXT j
      C5=C5+D5
      NEXT i
    B5=B5+(N-K)*C5*(1-Pk)^(K-5)
    NEXT K
E5=N-(B5*Pk^5)/2
```

**'Calculation for 6 TELs**

```
B6=0
FOR K=6 TO N-1
    C6=0
    FOR i1=1 TO K-5
      D6=0
      FOR i2=1 TO K-4-i1
        E6=0
        FOR i3=1 TO K-3-i1-i2
          E6=E6+(K-1-i1-i2-i3)*(K-2-i1-i2-i3)
          NEXT i3
        D6=D6+E6
        NEXT i2
      C6=C6+D6
      NEXT i1
    B6=B6+(N-K)*C6*(1-Pk)^(K-6)
    NEXT K
E6=N-(B6*Pk^6)/2
```

**'Output routines, print both to screen and to clipboard**

```
PRINT USING "###"; N,
PRINT USING "##.##"; Pk,
PRINT USING "###.###"; E1,E2,E3,E4,E5,E6,
PRINT USING "##.##"; E1*Pk, E2*Pk, E3*Pk, E4*Pk, E5*Pk, E6*Pk
z1(x)=N
z2(x)=Pk
z3(x)=E1
z4(x)=E2
```

```
z5(x)=E3
z6(x)=E4
z7(x)=E5
z8(x)=E6

WRITE #1, z1(x), z2(x), z3(x), z4(x), z5(x), z6(x), z7(x), z8(x)

'change value of Pk and repeat calculation
IF Pk>.2 THEN del=.1 ELSE del=.02
IF Pk<.06 THEN del=.01
Pk=Pk-del

WEND

'Request new inputs to run another case
INPUT "change number of TBMs"; s$
IF s$="y" GOTO 100
GOTO 110
100 INPUT "Number of TBMs:", N
GOTO 10

'End program
110 CLOSE #1
```

### 3. Markov model program to obtain expected-value outcomes for prelaunch counterforce and postlaunch counterbattery attacks

```
CLS
OPEN "clip:" FOR OUTPUT AS #1

'Input values for calculation
INPUT "Number of TELs="; NuTEL
INPUT "prelaunch prob of kill=";p1
INPUT "post-launch prob of KILL=";p2
INPUT "Maximum number of TBMs=";N
q=(1-p1)*(1-p2)
p=1-q

'Dimension arrays
DIM T(20,100)
```

```
DIM L(100)
DIM Tot(100)

'Set initial conditions, T is prob state of TEL set
FOR i=0 TO NuTEL-1
   T(i,0)=0
   NEXT i
T(NuTEL,0)=1
x=0

WHILE x<N:  'Calculate outcome for all x<=N
x=x+1
L(x)=(1-T(0,x-1))*(1-p1):  'L is probability of a TBM launch for each x
Tot(x)=Tot(x-1)+L(x):  'Tot is cumulative number of TBMs launched
     at x

'Matrix calculation for TEL prob state
T(NuTEL,x)=q*T(NuTEL,x-1)
FOR i=1 TO NuTEL-1
   T(i,x)=q*T(i,x-1)+p*T(i+1,x-1)
   NEXT i
T(0,x)=T(0,x-1)+p*T(1,x-1)

'Show outcomes for each x
PRINT x, L(x), Tot(x), T(0,x),
WRITE #1, x,L(x),Tot(x)
WEND

CLOSE #1
```

4. **Monte Carlo program for assessing outcomes of combined prelaunch counterforce and postlaunch counterbattery attacks on reducing the total number of TBMs launched.**

```
'Counterforce and Counterbattery, MC.4
```

```
'This progam does the following things:

'  1. In monte carlo mode calculates number of TBMs launched,
'     assuming pre- and post-launch counterforce attacks by the
      defended.
'  2. Runs in automatic mode 500 trials for each case.
'  3. Performs cases for input values for TBM and TEL inventories.
'  4. Performs cases for input values for pre- and post-launch
      counterforce effectiveness
'       (Pk1 and Pk2)
'  5. Calculates expected outcomes and their standard deviations.
'  6. Keeps statistics on outcomes per trial, prints distributions of
      outcomes.

'Start of program
CLS
INPUT "Specify name of output file "; z$
OPEN z$ FOR OUTPUT AS #1
DIM T(1000)
DIM Attack(1000)
DIM Outcome(1000)
DIM c(1000)
DIM d(1000)
DIM liveTEL(1000)

RANDOMIZE TIMER

'Inputs for case
INPUT "Total TBM inventory per cluster"; totTBM
INPUT "Total number of TELs per cluster"; totTEL
INPUT "Pre-launch Counterforce Effectiveness"; Pk1
INPUT "Post-launch Counterbattery Effectiveness"; Pk2
INPUT "Number of random trials desired, <Return> sets no. at 500";
    Tr
IF Tr=0 THEN Tr=500

'Prints (to readable file) input variables at the start of each case
PRINT #1, " "
PRINT #1, "Total TELs, TBMs in a cluster"; totTEL; totTBM,
PRINT #1, "Max Salvos"; W; "Pk1"; Pk,"Pk2",Pk2, "Num of trials";Tr
```

```
'Initializes variables at the start of each case
FOR y=0 TO 999
   c(y)=0
   d(y)=0
   NEXT y
x=0
umax=0
SumAttack=0
SumVar=0

'Performs calculations for TBMs launched for Tr trials
FOR x=1 TO Tr

'Set up initial conditions for calculations of each trial
liveTEL(x)=totTEL
NuTBM=totTBM
Attack(x)=0

'Checks to see whether calculation should stop
WHILE NuTBM*liveTEL(x)>0

'Pre-launch attack calculation
NuTBM=NuTBM-1
a=RND
IF a<Pk1 THEN
   liveTEL(x)=liveTEL(x)-1
   GOTO 70
END IF

'Accumulates total TBM launches
Attack(x)=Attack(x)+1

'Post-launch attack calculation
 a=RND
 IF a<Pk2 THEN liveTEL(x)=liveTEL(x)-1

70 WEND
```

```
'Summary data from trial
SumAttack=SumAttack+Attack(x)
SumVar=SumVar+Attack(x)^2

'Prepares histogram for total TBMs launched per trial
y=Attack(x)
c(y)=c(y)+1

'Prepares histogram for surviving TELs
z=liveTEL(x)
d(z)=d(z)+1

120 NEXT x

'Calculations for trials can take awhile, alerts user case done
BEEP

'Prepares summary and distributional output data for each case
Ave=SumAttack/Tr
Var=(SumVar/Tr)-Ave^2
Sigma=SQR(Var*Tr/(Tr-1))
Sig.mean=Sigma/SQR(Tr)

'Prints data to readable file
PRINT #1, "Trials=";Tr,"Ave No. TBMs fired=";Ave, "Std
    Dev=";Sigma,"Std Dev Mean=";Sig.mean
PRINT #1, " "
PRINT #1, "Histogram of number of TBMs fired"
PRINT #1, "    ";
PRINT #1, USING "#####";0;1;2;3;4;5;6;7;8;9
FOR j=0 TO totTBM STEP 10

PRINT #1, USING "#####";j;c(j);c(j+1);c(j+2);c(j+3);c(j+4);c(j+5);
c(j+6);c(j+7);c(j+8);c(j+9)
  NEXT j
PRINT #1, " "
PRINT #1, "Histogram of number of TELs surviving"
PRINT #1, "    ";
```

```
PRINT #1, USING "#####";0,1,2,3,4,5,6,7,8,9
FOR j=0 TO totTEL STEP 10

PRINT #1, USING"#####";j;d(j);d(j+1);d(j+2);d(j+3);d(j+4);d(j+5);
d(j+6);d(j+7);d(j+8);d(j+9)
   NEXT j
```

**'Closes the program**

```
CLOSE #1
END
```